Solutions Manual to Accompany

Introduction to Linear Regression Analysis

Sixth Edition

Solutions Manual to Accompany
Introduction to Linear Regression Analysis

Sixth Edition

Douglas C. Montgomery
Arizona State University
School of Computing, Informatics, and Decisions Systems Engineering
Tempe, AZ

Elizabeth A. Peck
The Coca-Cola Company (retired)
Atlanta, GA

G. Geoffrey Vining
Virginia Tech
Department of Statistics
Blacksburg, VA

Prepared by

Anne G. Ryan
Virginia Tech
Department of Statistics
Blacksburg, VA

This sixth edition first published 2022

© 2022 John Wiley and Sons, Inc.

Edition History
John Wiley and Sons, Inc. (5e, 2012)

Registered Offices
John Wiley & Sons, Inc., 111 River Street, Hoboken, NJ 07030, USA

Editorial Office
111 River Street, Hoboken, NJ 07030, USA

For details of our global editorial offices, customer services, and more information about Wiley products visit us at www.wiley.com.

Wiley also publishes its books in a variety of electronic formats and by print-on-demand. Some content that appears in standard print versions of this book may not be available in other formats.

Library of Congress Cataloging-in-Publication Data Applied for
ISBN 978-1-119-57869-7 (paper)

Cover Design: Wiley

Set in 10/12pt Computer Modern Roman by Straive, Chennai, India

CONTENTS

PREFACE

This book contains the complete solutions to the first eight chapters and the odd-numbered problems for chapters nine through fifteen in *Introduction to Linear Regression Analysis, Sixth Edition*. The solutions were obtained using Minitab®, JMP®, and SAS®.

The purpose of the solutions manual is to provide students with a reference to check their answers and to show the complete solution. Students are advised to try to work out the problems on their own before appealing to the solutions manual.

Anne R. Driscoll
Virginia Tech

Chapter 2: Simple Linear Regression

2.1   a. $\widehat{y} = 21.8 - .007x_8$

    b.

Source	d.f.	SS	MS
Regression	1	178.09	178.09
Error	26	148.87	5.73
Total	27	326.96	

    c. A 95% confidence interval for the slope parameter is $-0.007025 \pm 2.056(0.00126) = (-0.0096, -0.0044)$.

    d. $R^2 = 54.5\%$

    e. A 95% confidence interval on the mean number of games won if opponents' yards rushing is limited to 2000 yards is $7.738 \pm 2.056(.473) = (6.766, 8.711)$.

2.2   The fitted value is 9.14 and a 90% prediction interval on the number of games won if opponents' yards rushing is limited to 1800 yards is $(4.935, 13.351)$.

2.3   a. $\widehat{y} = 607 - 21.4x_4$

    b.

Source	d.f.	SS	MS
Regression	1	10579	10579
Error	27	4103	152
Total	28	14682	

    c. A 99% confidence interval for the slope parameter is $-21.402 \pm 2.771(2.565) = (-28.51, -14.29)$.

    d. $R^2 = 72.1\%$

    e. A 95% confidence interval on the mean heat flux when the radial deflection is 16.5 milliradians is $253.96 \pm 2.145(2.35) = (249.15, 258.78)$.

2.4   a. $\widehat{y} = 33.7 - .047x_1$

    b.

Source	d.f.	SS	MS
Regression	1	955.34	955.34
Error	30	282.20	9.41
Total	31	1237.54	

c. $R^2 = 77.2\%$

d. A 95% confidence interval on the mean gasoline mileage if the engine displacement is 275 in^3 is $20.685 \pm 2.042(.544) = (19.573, 21.796)$.

e. A 95% prediction interval on the mean gasoline mileage if the engine displacement is 275 in^3 is $20.685 \pm 2.042(3.116) = (14.322, 27.048)$.

f. Part d. is an interval estimator on the mean response at 275 in^3 while part e. is an interval estimator on a future observation at 275 in^3. The prediction interval is wider than the confidence interval on the mean because it depends on the error from the fitted model and the future observation.

2.5   a. $\widehat{y} = 40.9 - .00575x_{10}$

b.

Source	d.f.	SS	MS
Regression	1	921.53	921.53
Error	30	316.02	10.53
Total	31	1237.54	

c. $R^2 = 74.5\%$

The two variables seem to fit about the same. It does not appear that x_1 is a better regressor than x_{10}.

2.6   a. $\widehat{y} = 13.3 - 3.32x_1$

b.

Source	d.f.	SS	MS
Regression	1	636.16	636.16
Error	22	192.89	8.77
Total	23	829.05	

c. $R^2 = 76.7\%$

d. A 95% confidence interval on the slope parameter is $3.3244 \pm 2.074(.3903) = (2.51, 4.13)$.

e. A 95% confidence interval on the mean selling price of a house for which the current taxes are \$750 is $15.813 \pm 2.074(2.288) = (11.07, 20.56)$.

2.7   a. $\widehat{y} = 77.9 - 11.8x$

b. $t = \dfrac{11.8}{3.485} = 3.39$ with $p = 0.003$. The null hypothesis is rejected and we conclude there is a linear relationship between percent purity and percent of hydrocarbons.

c. $R^2 = 38.9\%$

    d. A 95% confidence interval on the slope parameter is $11.801 \pm 2.101(3.485)$ $= (4.48, 19.12)$.

    e. A 95% confidence interval on the mean purity when the hydrocarbon percentage is 1.00 is $89.664 \pm 2.101(1.025) = (87.51, 91.82)$.

2.8   a. $r = +\sqrt{R^2} = .624$

    b. This is the same as the test statistic for testing $\beta_1 = 0$, $t = 3.39$ with $p = 0.003$.

    c. A 95% confidence interval for ρ is

$$(\tanh[\text{arctanh}(.624) - 1.96/\sqrt{17}], \tanh[\text{arctanh}(.624) + 1.96/\sqrt{17}])$$
$$= \tanh(.267, 1.21) = (.261, .837)$$

2.9 The no-intercept model is $\widehat{y} = 2.414$ with MSE $= 21.029$. The MSE for the model containing the intercept is 17.484. Also, the test of $\beta_0 = 0$ is significant. Therefore, the model should not be forced through the origin.

2.10 a. $\widehat{y} = 69.104 + .419x$

    b. $r = .773$

    c. $t = 5.979$ with $p = 0.000$, reject H_0 and claim there is evidence that the correlation is different from zero.

    d. The test is

$$Z_0 = [\text{arctanh}(.773) - \text{arctanh}(.6)]\sqrt{26 - 3}$$
$$= (1.0277 - .6932)\sqrt{23}$$
$$= 1.60.$$

    Since the rejection region is $|Z_0| > Z_{\alpha/2} = 1.96$, we fail to reject H_0.

    e. A 95% confidence interval for ρ is

$$\tanh(1.0277 - (1.96)/\sqrt{23}) \le \rho \le \tanh(1.0277 + (1.96)/\sqrt{23}) = (.55, .89)$$

2.11 $\widehat{y} = .792x$ with MSE $= 158.707$. The model with the intercept has MSE $= 75.357$ and the test on β_0 is significant. The model with the intercept is superior.

2.12 a. $\widehat{y} = -6.33 + 9.21x$

    b. $F = 280590/4 = 74, 122.73$, it is significant.

    c. $H_0 : \beta_1 = 10000$ vs $H_1 : \beta_1 \ne 10000$ gives $t = (9.208 - 10)/.03382 = -23.4$ with $p = 0.000$. Reject H_0 and claim that the usage increase is less than 10,000.

    d. A 99% prediction interval on steam usage in a month with average ambient temperature of $58°$ is $527.759 \pm 3.169(2.063) = (521.22, 534.29)$.

2.13 a.

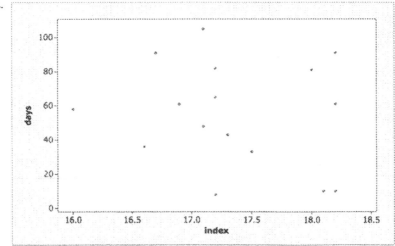

b. $\widehat{y} = 183.596 - 7.404x$

c. $F = 349.688/973.196 = .359$ with $p = 0.558$. The data suggests no linear association.

d.

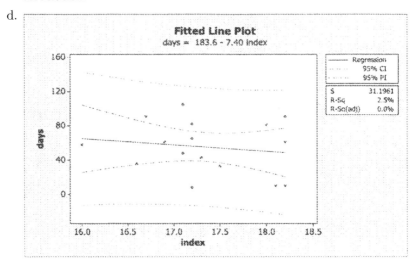

2.14 a.

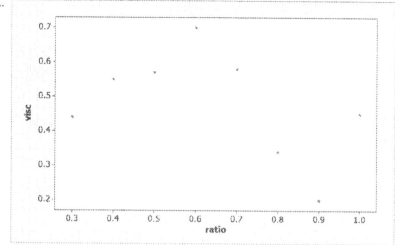

b. $\widehat{y} = .671 - .296x$

c. $F = .0369/.0225 = 1.64$ with $p = 0.248$. $R^2 = 21.5\%$. A linear association is not present.

d.

Fitted Line Plot
visc = 0.6714 - 0.2964 ratio

-------	Regression
-------	95% CI
· · · · ·	95% PI

S	0.149990
R-Sq	21.5%
R-Sq(adj)	8.4%

2.15   a. $\widehat{y} = 1.28 - .00876x$

c. $F = .32529..00225 = 144.58$ with $p = 0.000$. $R^2 = 96\%$. There is a linear association between viscosity and temperature.

c.

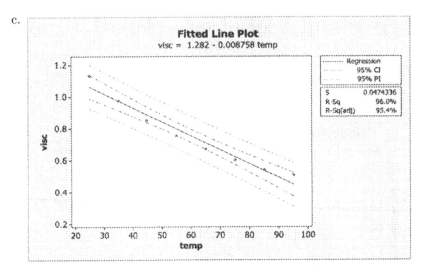

2.16 $\widehat{y} = -290.707 + 2.346x$, $F = 34286009$ with $p = 0.000$, $R^2 = 100\%$. There is almost a perfect linear fit of the data.

2.17 $\widehat{y} = 163.931 + 1.5796x$, $F = 226.4$ with $p = 0.000$, $R^2 = 93.8\%$. The model is a good fit of the data.

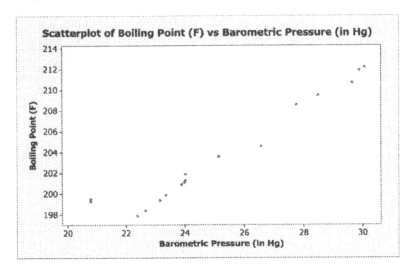

2.18 a. $\widehat{y} = 22.163 + 0.36317x$

  b. $F = 13.98$ with $p = 0.001$, so the relationship is statistically significant. However, the $R^2 = 42.4\%$, so there is still a lot of unexplained variation in this model.

c.

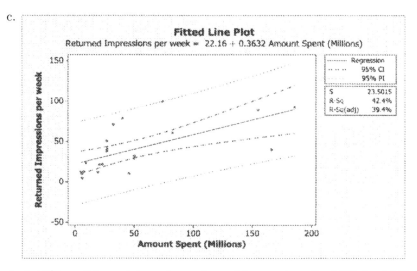

Fitted Line Plot
Returned Impressions per week = 22.16 + 0.3632 Amount Spent (Millions)

d. A 95% confidence interval on returned impressions for MCI (x=26.9) is

$$31.93 \pm (2.093)\sqrt{(552.3)\left(\frac{1}{21} + \frac{(26.9 - 50.4)^2}{111899}\right)} = (20.654, 43.206).$$

A 95% prediction interval is

$$31.93 \pm (2.093)\sqrt{(552.32)\left(1 + \frac{1}{21} + \frac{(26.9 - 50.4)^2}{111899}\right)} = (-18.535, 82.395).$$

2.19 a. $\widehat{y} = 130.2 - 1.249x$, $F = 72.09$ with $p = 0.000$, $R^2 = 75.8\%$. The model is a good fit of the data.

b. The fit for the SLR model relating satisfaction to age is much better compared to the fit for the SLR model relating satisfaction to severity in terms of R^2. For the SLR with satisfaction and age $R^2 = 75.8\%$ compared to $R^2 = 42.7\%$ for the model relating satisfaction and severity.

2.20 $\widehat{y} = 410.7 - 0.2638x$, $F = 7.51$ with $p = 0.016$, $R^2 = 34.9\%$. The engineer is correct that there is a relationship between initial boiling point of the fuel and fuel consumption. However, the $R^2 = 34.9\%$ indicating there is still a lot of unexplained variation in this model.

2.21 $\widehat{y} = 16.56 - 0.01276x$, $F = 4.94$ with $p = 0.034$, $R^2 = 14.1\%$. The winemaker is correct that sulfur content has a significant negative impact on taste with a $p - value = 0.034$. However, the $R^2 = 14.1\%$ indicating there is still a lot of unexplained variation in this model.

2.22 $\hat{y} = 21.25 + 7.80x$, $F = 0.22$ with $p = 0.648$, $R^2 = 1.3\%$. The chemist's belief
is incorrect. There is no relationship between the ratio of inlet oxygen to inlet
methanol and percent conversion ($p - value = 0.648$). The $R^2 = 1.3\%$, which
indicates that the ratio explains virtually none of the percent conversion.

2.23 a.

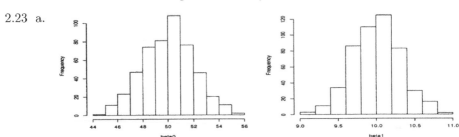

Both histograms are bell-shaped. The one for β_0 is centered around 50
and the one for β_1 is centered around 10.

b. The histogram is bell-shaped with a center of 100.

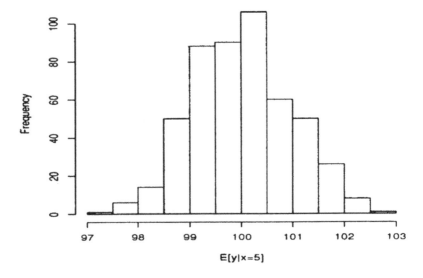

c. 481 out of 500 which is 96.2% which is very close to the stated 95%.

d. 474 out of 500 which is 94.8% which is very close to the states 95%.

2.24 Using a smaller value of n makes the estimates of the coefficients in the
regression model less precise. It also increases the variability in the predicted
value of y at $x = 5$. The lengths of the confidence intervals are wider for $n = 10$
and the histograms are more spread out.

2.25 a.

$$Cov(\widehat{\beta}_0, \widehat{\beta}_1) = Cov(\overline{y} - \widehat{\beta}_1 \overline{x}, \widehat{\beta}_1)$$

$$= Cov(\overline{y}, \widehat{\beta}_1) - \overline{x} Cov(\widehat{\beta}_1, \widehat{\beta}_1)$$

$$= 0 - \overline{x} \frac{\sigma^2}{S_{XX}} \quad \text{(by part b)}$$

$$= \frac{-\overline{x}\sigma^2}{S_{XX}}$$

b.

$$Cov(\overline{y}, \widehat{\beta}_1) = \frac{1}{nS_{XX}} Cov\left(\sum y_i, \sum (x_i - \overline{x})y_i\right)$$

$$= \frac{1}{nS_{XX}} \sum (x_i - \overline{x}) Cov(y_i, y_i)$$

$$= \frac{\sigma^2}{nS_{XX}} \sum (x_i - \overline{x})$$

$$= 0$$

2.26 a. Use the fact that $\dfrac{\text{SSE}}{\sigma^2} \sim \chi^2_{n-2}$. Then

$$E(\text{MSE}) = E\left(\frac{\text{SSE}}{n-2}\right)$$

$$= \frac{\sigma^2}{n-2} E(\chi^2_{n-2})$$

$$= \sigma^2$$

b. Use SSR $= \widehat{\beta}_1 S_{xy} = \widehat{\beta}_1^2 S_{xx}$.

$$E(\text{SSR}) = S_{xx} E(\widehat{\beta}_1^2)$$

$$= S_{xx}[Var(\widehat{\beta}_1 + (E(\widehat{\beta}_1))^2]$$

$$= S_{xx}\left(\frac{\sigma^2}{S_{xx}} + \beta_1^2\right)$$

$$= \sigma^2 + \beta_1^2 S_{xx}$$

2.27 a. No,

$$E(\widehat{\beta}_1) = E\left(\frac{\sum(x_i - \overline{x})y_i}{S_{xx}}\right)$$

$$= \frac{\sum(x_{i1} - \overline{x})}{S_{xx}}E(y_i)$$

$$= \frac{\sum(x_{i1} - \overline{x})}{S_{xx}}(\beta_0 + \beta_1 x_{i1} + \beta_2 x_{i2})$$

$$= \beta_1 + \frac{\sum(x_{i1} - \overline{x})x_{i2}}{S_{xx}}$$

b. The bias is

$$\beta_1 - E(\widehat{\beta}_1) = \frac{-\sum(x_{i1} - \overline{x})x_{i2}}{S_{xx}}$$

2.28 a. $\widetilde{\sigma}^2 = \text{SSE}/n$. So, $E(\widetilde{\sigma}^2) = \dfrac{n-2}{n}\sigma^2$ so the bias is $\left(1 - \dfrac{n-2}{n}\right)\sigma^2$.

b. As n gets large, the bias goes to zero.

2.29 If n is even, then half the points should be at $x = -1$ and the other half at $x = 1$. If n is odd, then one point should be at $x = 0$, then the rest of the points are evenly split between $x = -1$ and $x = 1$. There would be no way to test the adequacy of the model.

2.30 a. $r = +\sqrt{R^2} = 1.00$

b. The test of $\rho = 0$ is equivalent to the test of $\beta_1 = 0$. Therefore, $t = 272.25$ with $p = 0.000$.

c. For $H_0 : \rho = .5$, we get

$$Z_0 = [\text{arctanh}(.99) - \text{arctanh}(.5)]\sqrt{9}$$

$$= [2.647 - .549](3)$$

$$= 6.29.$$

We reject H_0.

d. $(\tanh[\text{arctanh}(.99) - 1.96/\sqrt{9}], \tanh[\text{arctanh}(.99) + 1.96/\sqrt{9}]) = (.963, .997)$

2.31 Since $R^2 = SS_R/S_{yy}$ and $S_{yy} = SS_R + SS_E$, then we need to show that in this case $SS_E > 0$. Now $SS_E = \sum(y_i - \widehat{y}_i)^2$, so for two different y_i's (say y_{1i} and y_{2i}) at the same value of x_i, both y_{1i} and y_{2i} cannot equal $\widehat{y}_i$ at x_i. Therefore at least one of $(y_{1i} - \widehat{y}_i)^2$ and $(y_{2i} - \widehat{y}_i)^2$ is > 0. Hence, $SS_E > 0$ and thus $R^2 < 1$.

2.32 a. $S(\beta_0, \beta_1) = \sum(y_i - \beta_0 - \beta_1 x_i)^2$ with β_0 known. We need to take the derivative of this with respect to β_1 and set it equal to zero. This gives

$$-2\sum_{i=1}^{n}(y_i - \beta_0 - \widehat{\beta}_1 x_i)x_i = 0$$

$$\widehat{\beta}_1 \sum_{i=1}^{n} x_i^2 = \sum_{i=1}^{n}(y_i - \beta_0)x_i$$

$$\widehat{\beta}_1 = \frac{\sum_{i=1}^{n}(y_i - \beta_0)x_i}{\sum_{i=1}^{n} x_i^2}$$

b.

$$Var(\hat{\beta}_1) = \frac{1}{\left(\sum_{i=1}^{n} x_i^2\right)^2} Var\left(\sum_{i=1}^{n} y_i x_i\right)$$

$$= \frac{1}{\left(\sum_{i=1}^{n} x_i^2\right)^2} \left(\sum_{i=1}^{n} x_i^2\right) \sigma^2$$

$$= \frac{\sigma^2}{\sum_{i=1}^{n} x_i^2}$$

c. $\dfrac{\widehat{\beta}_1 - \beta_1}{\sqrt{MS_E / \sum x_i^2}} \sim t_{n-2}$ so we get $\widehat{\beta}_1 \pm t_{\alpha/2, n-2}\sqrt{MS_E / \sum x_i^2}$ which is narrower than when both are unknown.

2.33

$$Var(e_i) = Var(y_i - \hat{y}_i)$$

$$= Var(y_i) + Var(\hat{y}_i) - 2Cov(y_i, \hat{y}_i)$$

$$= \sigma^2 + \left[\frac{\sigma^2}{n} + \frac{(x_i - \bar{x})^2 \sigma^2}{S_{xx}}\right] - 2\left[\frac{\sigma^2}{n} + \frac{(x_i - \bar{x})^2 \sigma^2}{S_{xx}}\right]$$

$$= \sigma^2 \left[1 - \frac{1}{n} - \frac{(x_i - \bar{x})^2}{S_{xx}}\right]$$

which depends on the value of x_i and thus is not constant.

2.34 a. $r = -0.7961$

   b. The test of $\rho = 0$ is equivalent to the test of $\beta_1 = 0$. Therefore, $t = -6.96$ with $p = 0.000$

   c. For H_0: $\rho = 0.5$, we get $Z_0 = [\mathrm{arctanh}(-0.7961) - \mathrm{arctanh}(0.5)] * \sqrt{27} = [-1.088 - 0.549] * \sqrt{27} = -8.51$. We reject H_0.

   d. $\left(\tanh\left[\mathrm{arctanh}(-0.7961) - \dfrac{1.96}{\sqrt{27}}\right], \tanh\left[\mathrm{arctanh}(-0.7961) + \dfrac{1.96}{\sqrt{27}}\right]\right) = (-0.8986, -0.6111)$

2.35 $\widehat{y} = 10.89 + 0.097x$, $F = 11.60$ with $p = 0.002$, $R^2 = 0.29$. The model using ERA is a superior model compared to the model using total runs scored to predicted team wins. The R^2 for the ERA model is 0.63 compared to the much smaller value of 0.29 for the total runs model. Also, the F ratio is much larger for the ERA model (48.46) compared to the model using total runs scored (11.60).

2.36 a. $F = 421.02$, $p =< 0.001$. We reject H_0 and conclude that there is significant relationship between the median price per square foot and the median home rental price.

   b. [5.43, 6.62]

   c. $R^2 = 0.90$. Yes, the median price per square foot explains about 90% of the variability in the median home rental price.

2.37 a. $r = 0.9464$

   b. The test of $\rho = 0$ is equivalent to the test of $\beta_1 = 0$. Therefore, $t = 20.52$ with $p =< 0.0001$

   c. For H_0: $\rho = 0.5$, we get $Z_0 = [\mathrm{arctanh}(0.9464) - \mathrm{arctanh}(0.5)] * \sqrt{(51-3)} = [1.796 - 0.549] * \sqrt{(51-3)} = 8.64$. We reject H_0.

   d. $\left(\tanh\left[\mathrm{arctanh}(0.9464) - \dfrac{1.96}{\sqrt{(51-3)}}\right], \tanh\left[\mathrm{arctanh}(0.9464) + \dfrac{1.96}{\sqrt{(51-3)}}\right]\right) = (0.908, 0.969)$

2.38 e. None of the above

2.39 b. 0.025

2.40 a. True

2.42 a. True

2.43 b. False

2.44 a. True

2.45 a. True

Chapter 3: Multiple Linear Regression

3.1 a. $\widehat{y} = -1.8 + .0036x_2 + .194x_7 - .0048x_8$.

b. Regression is significant.

Source	d.f.	SS	MS	F	p-value
Regression	3	257.094	85.698	29.44	0.000
Error	24	69.87	2.911		
Total	27	326.964			

c. All three are significant.

Coefficient	test statistic	p-value
β_2	5.18	0.000
β_7	2.20	0.038
β_8	-3.77	0.001

d. $R^2 = 78.6\%$ and $R^2_{Adj} = 76.0\%$

e. $F_0 = (257.094 - 243.03)/2.911 = 4.84$ which is significant at $\alpha = 0.05$. The test statistic here is the square of the t-statistic in part c.

3.2 Correlation coefficient between y_i and $\widehat{y}_i$ is .887. So $(.887)^2 = .786$ which is R^2.

3.3 a. A 95% confidence interval on the slope parameter β_7 is $\widehat{\beta}_7 \pm 2.064(.08823) = (.012, .376)$

b. A 95% confidence interval on the mean number of games won by a team when $x_2 = 2300$, $x_7 = 56.0$ and $x_8 = 2100$ is

$$\widehat{y} \pm t_{\alpha/2,24}\sqrt{\widehat{\sigma}\mathbf{x}_0'(\mathbf{X}'\mathbf{X})^{-1}\mathbf{x}_0} = 7.216 \pm 2.064(.378)$$

$$= (6.44, 7.99)$$

3.4 a. $\widehat{y} = 17.9 + .048x_7 - .00654x_8$ with $F = 15.13$ and $p = 0.000$ which is significant.

b. $R^2 = 54.8\%$ and $R^2_{Adj} = 51.5\%$ which are much lower.

c. For β_7, a 95% confidence interval is $0.484 \pm 2.064(.1192) = (-.198, .294)$ and for the mean number of games won by a team when $x_7 = 56.0$ and $x_8 = 2100$, a 95% confidence interval is $6.926 \pm 2.064(.533) = (5.829, 8.024)$. Both lengths are greater than when x_2 was included in the model.

d. It can affect many things including the estimates and standard errors of the coefficients and the value of R^2.

13

3.5   a. $\hat{y} = 32.9 - .053x_1 + .959x_6$

b. Regression is significant.

Source	d.f.	SS	MS	F	p-value
Regression	2	972.9	486.45	53.31	0.000
Error	29	264.65	9.13		
Total	31	1237.54			

c. $R^2 = 78.6\%$ and $R^2_{Adj} = 77.3\%$. For the simple linear regression with x_1, $R^2 = 77.2\%$.

d. A 95% confidence interval for the slope parameter β_1 is $-.053 \pm 2.045(.006145) = (-.0656, -.0405)$.

e. x_1 is significant while x_6 is not.

Coefficient	test statistic	p-value
β_1	-8.66	0.000
β_6	1.43	0.163

f. A 95% confidence interval on the mean gasoline mileage when $x_1 = 275$ in^3 and $x_6 = 2$ is $20.187 \pm 2.045(.643) = (18.872, 21.503)$.

g. A 95% prediction interval for a new observation on gasoline mileage when $x_1 = 275$ in^3 and $x_6 = 2$ is $20.187 \pm 2.045(3.089) = (13.887, 26.488)$

3.6   The lengths from problem 2.4 are 2.223 and 12.716, respectively. For problem 3.5, they are 2.631 and 12.634. The lengths are pretty much the same which indicates that adding x_6 does not help much.

3.7   a. $\hat{y} = 14.9 + 1.92x_1 + 7.00x_2 + .149x_3 + 2.72x_4 + 2.01x_5 - .41x_6 - 1.4x_7$

$- .0371x_8 + 1.56x_9$

b. $F = 9.04$ with $p = 0.000$ which is significant.

c. None of the t-tests are significant. There is a multicollinearity problem.

d. $F = \dfrac{(707.298 - 701.69)/2}{8.696} = .322$ which indicates their is no contribution of lot size and living space given that all the other regressors are in the model.

e. Yes, there is a multicollinearity problem.

3.8   a. $\hat{y} = 2.53 + .0185x_6 + 2.19x_7$

b. $F = 27.95$ with $p = 0.000$ which is significant. $R^2 = 70.0\%$ and $R^2_{Adj} = 67.5\%$.

c. Both are significant.

Coefficient	test statistic	p-value
β_6	6.74	0.000
β_7	2.25	0.034

d. For β_6, a 95% confidence interval is $.0185 \pm 2.064(.0027) = (.013, .024)$ and for β_7, a 95% confidence interval is $2.185 \pm 2.064(.9727) = (.177, 4.193)$.

e. $t = 6.62$ with $p = 0.000$ which is significant. $R^2 = 63.6\%$ and $R^2_{Adj} = 62.2\%$. These are basically the same as in part b.

f. A 95% confidence interval on the slope parameter β_6 is $.019 \pm 2.064(.0029) = (.013, .025)$. The length of this confidence interval is almost exactly the same as the one from the model including x_7.

g. As always, MS_{Res} is lower when x_6 and x_7 are in the model.

3.9 a. $\hat{y} = .00483 - .345x_1 - .00014x_4$

b. $F = 24.66$ with $p = 0.000$ which is significant.

c. $R^2 = 66.4\%$ and $R^2_{Adj} = 63.7\%$

d. x_1 is significant while x_4 is not.

Coefficient	test statistic	p-value
β_1	-5.12	0.000
β_7	$-.02$	0.986

e. It doesn't appear to be.

3.10 a. $\hat{y} = 4.00 + 2.34x_1 + .403x_2 + .273x_3 + 1.17x_4 - .684x_5$

b. $F = 16.51$ with $p = 0.000$ which is significant.

c. x_4 and x_5 appear to contribute to the model.

Coefficient	test statistic	p-value
β_1	1.35	0.187
β_2	1.77	0.086
β_3	0.82	0.418
β_4	3.84	0.001
β_5	-2.52	0.017

d. For the model in part a, $R^2 = 72.1\%$ and $R^2_{Adj} = 67.7\%$. For the model with only aroma and flavor, $R^2 = 65.9\%$ and $R^2_{Adj} = 63.9\%$. These are basically the same.

e. For the model in part a, the confidence interval is 1.1683 $\pm$ 2.0369(.3045) = (.548, 1.789). For the model with only aroma and flavor, the confidence interval is 1.1702 $\pm$ 2.0301(.2905) = (.581, 1.759). These two intervals are almost the same.

3.11 a. $\widehat{y} = 32.1 + .0556x_1 + .282x_2 + .125x_3 - .000x_4 - 16.1x_5$

b. $F = 29.86$ with $p = 0.000$ which is significant.

c. x_2 and x_5 appear to contribute to the model.

Coefficient	test statistic	p-value
β_1	1.86	0.093
β_2	4.90	0.001
β_3	0.31	0.763
β_4	-0.00	1.00
β_5	-11.03	0.000

d. For the model in part a, $R^2 = 93.7\%$ and $R^2_{Adj} = 90.6\%$. For the model with only temperature and particle size, $R^2 = 91.5\%$ and $R^2_{Adj} = 90.2\%$. These are basically the same.

e. For the model in part a, a 95% confidence interval is .282 $\pm$ 2.228(.05761) = (.154, .410). For the model with only aroma and flavor, a 95% confidence interval is .282 $\pm$ 2.16(.05883) = (.155, .409). These two intervals are almost the same.

3.12 a. $\widehat{y} = 11.1 + 350x_1 + .109x_2$

b. $F = 87.6$ with $p = 0.000$ which is significant.

c. Both contribute to the model.

Coefficient	test statistic	p-value
β_1	8.82	0.000
β_2	10.91	0.000

d. For the model in part a, $R^2 = 84.2\%$ and $R^2_{Adj} = 83.2\%$. For the model with only time, $R^2 = 46.8\%$ and $R^2_{Adj} = 45.2\%$. These are very different and suggest that amount of surfactant is needed in the model.

e. For the model in part a, a 95% confidence interval is .1089 $\pm$ 2.0345(.00998) = (.089, .129). For the model with only time, a 95% confidence interval is .0977 $\pm$ 2.0322(.01788) = (.061, .134). These second interval is wider.

3.13 a. $\widehat{y} = 5.89 - .498x_1 + .183x_2 + 35.4x_3 + 5.84x_4$

b. $F = 31.92$ with $p = 0.000$ which is significant.

c. x_2 and x_3 contribute to the model.

Coefficient	test statistic	p-value
β_1	1.41	0.165
β_2	10.63	0.000
β_3	3.19	0.002
β_4	2.01	.049

d. For the model in part a, $R^2 = 69.1\%$ and $R^2_{Adj} = 67.0\%$. For the model with only x_2 and x_3, $R^2 = 66.6\%$ and $R^2_{Adj} = 65.5\%$. These are basically the same.

e. For the model in part a, a 99% confidence interval is $.1827 \pm 2(.01718) = (.148, .217)$. For the model with only x_2 and x_3, a 99% confidence interval is $.1846 \pm 2(.01755) = (.149, .219)$. These intervals are basically the same.

3.14 a. $\widehat{y} = .679 + 1.41x_1 - .0156x_2$

b. $F = 85.46$ with $p = 0.000$ which is significant.

c. Both contribute to the model.

Coefficient	test statistic	p-value
β_1	7.15	0.000
β_2	−10.95	0.000

d. For the model in part a, $R^2 = 82.2\%$ and $R^2_{Adj} = 81.2\%$. For the model with only temperature, $R^2 = 57.6\%$ and $R^2_{Adj} = 56.5\%$. These are very different and suggest that the ratio variable is needed in the model.

e. For the model in part a, a 99% confidence interval is $-.0156 \pm 2.7(.0014) = (-.019, -.012)$. For the model with only time, a 99% confidence interval is $-.0156 \pm 2.7(.0022) = (-.022, -.009)$. The second interval is wider.

3.15 a. $\widehat{y} = 996 + 1.41x_1 - 14.8x_2 + 3.20x_3 - 0.108x_4 + 0.355x_5$

b. $F = 22.39$ with $p = 0.000$ which is significant.

c. PRECIP(x_1), EDUC(x_2), NONWHITE(x_3), and SO2(x_5) contribute to the model.

Coefficient	test statistic	p-value
β_1	2.04	0.046
β_2	−2.11	0.040
β_3	5.14	0.000
β_4	−0.80	0.427
β_5	3.90	0.000

d. $R^2 = 67.5\%$ and $R^2_{Adj} = 64.4\%$.

e. A 95% confidence interval on β_5 is $0.355 \pm (2.005)(0.09096) = (0.1726, 0.5374)$

3.16 a. For LifeExp, $\widehat{y} = 70.2 - 0.0226x_1 - 0.000447x_2$.
   For LifeExpMale, $\widehat{y} = 73.1 - 0.0257x_1 - 0.000479x_2$.
   For LifeExpFemale, $\widehat{y} = 67.4 - 0.0199x_1 - 0.000409x_2$.

b. For LifeExp, $F = 13.46$ with $p = 0.000$ which is significant.
   For LifeExpMale, $F = 12.53$ with $p = 0.000$ which is significant.
   For LifeExpFemale, $F = 14.07$ with $p = 0.000$ which is significant.

c. Both predictors are significant in all three models.

Model	Coefficient	test statistic	p-value
LifeExp	β_1	-2.35	0.024
LifeExp	β_2	-2.22	0.033
LifeExpMale	β_1	-2.34	0.025
LifeExpMale	β_2	-2.07	0.046
LifeExpFemale	β_1	-2.36	0.024
LifeExpFemale	β_2	-2.31	0.027

d. For LifeExp, $R^2 = 43.5\%$, $R^2_{Adj} = 40.2\%$.
   For LifeExpMale, $R^2 = 41.7\%$, $R^2_{Adj} = 38.4\%$.
   For LifeExpFemale, $R^2 = 44.6\%$, $R^2_{Adj} = 41.4\%$.

e. For LifeExp, $-0.0004470 \pm (2.024)(0.0002016) = (-0.000855, -.00003896)$.
   For LifeExpMale, $-0.0004785 \pm (2.024)(0.0002308) = (-0.0009456, -0.00001136)$.
   For LifeExpFemale, $-0.0004086 \pm (2.024)(0.0001766) = (-0.000766, -0.00005116)$.

3.17 The multiple linear regression model that relates age, severity, and anxiety to patient satisfaction is significant with $F = 30.97$ and $p = 0.000$. It also appears that age and severity contribute significantly to the model, while anxiety is insignificant ($p = 0.417$). Compared to the simple linear regression in Section 2.7 that related only severity to patient satisfaction, the addition of age and anxiety has improved the model. The R^2 has increased from 0.43 to 0.82. The mean square error in the multiple linear regression is 95.1, considerably smaller than the MSE in the simple linear regression, which was 270.02. Compared to the multiple linear regression is Section 3.6, adding anxiety to the model does not seem to improve the model. The R^2_{Adj} decreases slightly from 0.792 to 0.789, the MSE increases from 93.7 to 95.1, and the regressor is insignificant with $p = 0.417$.
The regression equation is $\widehat{y} = 140 - 1.12x_{age} - 0.463x_{severity} + 1.21x_{anxiety}$.

Coefficient	test statistic	p-value
β_{age}	-6.11	0.000
$\beta_{severity}$	-2.53	0.019
$\beta_{anxiety}$	0.83	0.417

3.18 The multiple linear regression model for the fuel consumption data is insignificant with $F = 0.94$ and $p = 0.527$. The variance inflation factors (VIFs) indicate a severe multicollinearity problem with many VIFs much greater than 10. In addition none of the t-tests are significant. This model is not satisfactory.

The regression equation is $\widehat{y} = -315 + 0.159x_2 + 1.03x_3 - 8.6x_4 - 0.432x_5 - 0.14x_6 - 0.32x_7 - 0.52x_8$.

Coefficient	test statistic	p-value	VIF
β_2	0.17	0.871	1.901
β_3	0.36	0.729	168.467
β_4	-0.19	0.851	43.104
β_5	-0.47	0.648	60.791
β_6	-0.12	0.910	275.473
β_7	-0.10	0.924	185.707
β_8	-0.24	0.819	44.363

3.19 The multiple linear regression model for the wine quality of young red wines is significant with $F = 6.25$ and $p = 0.000$. However, x_7 the anthocyanin color and x_{10} the ionized anthocyanins (percent) are removed from the model due to linear dependencies. The anthocyanin color is equal to the wine color minus polymeric pigment color $(x_5 - x_6)$. The ionized anthocyanins is equal to $\dfrac{x_5 - x_6}{50}$.

The VIFs indicate an extreme problem with multicollinearity. Remedial methods will be discussed in Chapter 9. Due to multicollinearity caution is taken when making interpretations from this model.

The regression equation is $\widehat{y} = -5.2 + 6.15x_2 + 0.00455x_3 - 2.96x_4 + 6.58x_5 - 0.66x_6 - 14.5x_8 - 0.261x_9$.

Coefficient	test statistic	p-value	VIF
β_2	1.77	0.090	3.834
β_3	0.59	0.560	3.482
β_4	-1.37	0.183	543.612
β_5	2.15	0.042	444.590
β_6	-0.37	0.711	30.433
β_8	-1.87	0.074	7.356
β_9	-1.32	0.200	27.849

3.20 The multiple linear regression model for methanol oxidation data is significant with $F = 28.02$ and $p = 0.000$. The $R^2 = 92.1\%$ and $R^2_{Adj} = 88.8\%$. The variables x_1, x_2 and x_3 seem to contribute to the model based on the t-tests, however there is a problem with multicollinearity as evident by the VIFs. Due to multicollinearity caution is taken when making interpretations from this model.

The regression equation is $\widehat{y} = -2669 + 22.3x_1 + 3.89x_2 + +102x_3 + 0.81x_4 - 1.63x_5$.

Coefficient	test statistic	p-value	VIF
β_1	3.09	0.009	1.519
β_2	5.70	0.000	26.284
β_3	3.91	0.002	26.447
β_4	0.21	0.840	2.202
β_5	-0.21	0.833	1.923

3.21 a. If $x_2 = 2$, then for model (1), $\widehat{y} = 108 + .2x_1$ and for model (2), $\widehat{y} = 101 + 2.15x_1$. If $x_2 = 8$, then for model (1), $\widehat{y} = 132 + .2x_1$ and for model (2), $\widehat{y} = 119 + 8.15x_1$. The interaction term in model 2 affects the slope of the line.

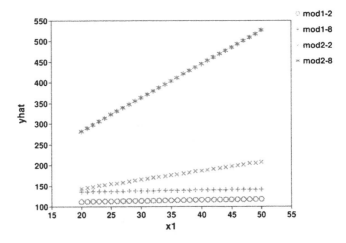

b. This is just the slope which is .2 regardless of the value of x_2.

c. The mean change here is $5 + .15$ which is $x_2 + .15$. Thus the result depends on the value of x_2.

3.22

$$F = \frac{MS_R}{MS_E}$$

$$= \frac{SS_R/k}{SS_E/(n-k-1)}$$

$$= \frac{SS_R/(p-1)}{SS_E/(n-p)}$$

$$= \frac{SS_R/(p-1)(S_{yy})}{SS_E/(n-p)(S_{yy})}$$

$$= \frac{R^2(n-p)}{(p-1)(1-R^2)}$$

$$= F_0$$

which then has an F distribution with $p-1$ and $n-p$ degrees of freedom.

3.23 a. $F_0 = \dfrac{(.9)(25-3)}{(3-1)(1-.9)} = 99$ which exceeds the critical value of $F_{.05,2,22} = 3.44$ so H_0 is rejected.

b. The value of R^2 should be surprisingly low.

$$\frac{R^2(n-p)}{(p-1)(1-R^2)} > 3.44$$

$$\frac{R^2(22)}{(2)(1-R^2)} > 3.44$$

$$\frac{R^2}{1-R^2} > .312727$$

$$R^2 > .312727 - .312727R^2$$

$$R^2 > .238$$

3.24 $SS_R = \widehat{\beta}'\mathbf{X}'\mathbf{y} - \dfrac{\left(\sum y_i\right)^2}{n} = \mathbf{y}'\mathbf{X}(\mathbf{X}'\mathbf{X})^{-1}\mathbf{X}'\mathbf{y} - \dfrac{n^2\overline{y}^2}{n} = \sum_{i=1}^{n}\widehat{y}_i^2 - n\overline{y}^2$

3.25 a. Use $\mathbf{T}\beta = \mathbf{c}$.

$$\mathbf{T} = \begin{pmatrix} 0 & 1 & -1 & 0 & 0 \\ 0 & 0 & 1 & -1 & 0 \\ 0 & 0 & 0 & 1 & -1 \end{pmatrix} \quad \beta = \begin{pmatrix} \beta_0 \\ \beta_1 \\ \beta_2 \\ \beta_3 \\ \beta_4 \end{pmatrix} \quad \mathbf{c} = \begin{pmatrix} 0 \\ \beta \\ \beta \\ \beta \end{pmatrix}$$

b. Use β from part a, $\mathbf{c} = \begin{pmatrix} 0 \\ 0 \end{pmatrix}$ and

$$\mathbf{T} = \begin{pmatrix} 0 & 1 & -1 & 0 & 0 \\ 0 & 0 & 0 & 1 & -1 \end{pmatrix}$$

c. Use β from part a, $\mathbf{c} = \begin{pmatrix} 0 \\ 0 \end{pmatrix}$ and

$$\mathbf{T} = \begin{pmatrix} 0 & 1 & -2 & -4 & 0 \\ 0 & 1 & 2 & 0 & 0 \end{pmatrix}$$

3.26 a. Consider a new variable $z = \begin{Bmatrix} 0 & \text{if sample 1} \\ 1 & \text{if sample 2} \end{Bmatrix}$. Then write the model as
$y_i = \beta_0 + \beta_1 x_i + (\gamma_0 - \beta_0)z + (\gamma_1 - \beta_1)x_i z + \varepsilon_i$.

b. Call $\gamma_0 - \beta_0 = \nu_1$ and $\gamma_1 - \beta_1 = \nu_2$. Then we want to test $H0 : \nu_2 = 0$.
Then use

$$\mathbf{T} = (0\ 0\ 0\ 1) \quad \boldsymbol{\beta} = \begin{pmatrix} \beta_0 \\ \beta_1 \\ \nu_1 \\ \nu_2 \end{pmatrix} \quad c = 0$$

c. This is test of $\nu_1 = 0$ and $\nu_2 = 0$.

$$\mathbf{T} = \begin{pmatrix} 0 & 0 & 1 & 0 \\ 0 & 0 & 0 & 1 \end{pmatrix} \quad \boldsymbol{\beta} = \begin{pmatrix} \beta_0 \\ \beta_1 \\ \nu_1 \\ \nu_2 \end{pmatrix} \quad \mathbf{c} = \begin{pmatrix} 0 \\ 0 \end{pmatrix}$$

d. Use $\beta_1 = c$ and $\nu_2 = 0$.

$$\mathbf{T} = \begin{pmatrix} 0 & 1 & 0 & 0 \\ 0 & 0 & 0 & 1 \end{pmatrix} \quad \boldsymbol{\beta} = \begin{pmatrix} \beta_0 \\ \beta_1 \\ \nu_1 \\ \nu_2 \end{pmatrix} \quad \mathbf{c} = \begin{pmatrix} c \\ 0 \end{pmatrix}$$

3.27

$$\begin{aligned} Var(\widehat{\mathbf{y}}) &= Var(\mathbf{X}\widehat{\boldsymbol{\beta}}) \\ &= \mathbf{X}[Var(\widehat{\boldsymbol{\beta}})]\mathbf{X}' \\ &= [\mathbf{X}(\mathbf{X}'\mathbf{X})^{-1}\mathbf{X}']\sigma^2 \\ &= \sigma^2 \mathbf{H} \end{aligned}$$

3.28

$$\begin{aligned} \mathbf{HH} &= \mathbf{X}(\mathbf{X}'\mathbf{X})^{-1}\mathbf{X}'\mathbf{X}(\mathbf{X}'\mathbf{X})^{-1}\mathbf{X}' \\ &= \mathbf{X}(\mathbf{X}'\mathbf{X})^{-1}\mathbf{X}' \\ &= \mathbf{H} \end{aligned}$$

and

$$(\mathbf{I} - \mathbf{H})(\mathbf{I} - \mathbf{H}) = \mathbf{I} - \mathbf{H} - \mathbf{H} + \mathbf{H}\mathbf{H}$$
$$= \mathbf{I} - \mathbf{H} - \mathbf{H} + \mathbf{H}$$
$$= \mathbf{I} - \mathbf{H}$$

3.29 First note that $(\mathbf{X}'\mathbf{X})^{-1} = \dfrac{1}{S_{xx}}\begin{pmatrix} \sum x_i^2/n & -\bar{x} \\ -\bar{x} & 1 \end{pmatrix}$. When x_i moves further from $\bar{x}$, both h_{ii} and h_{ij} increase.

$$h_{ii} = \begin{pmatrix} 1 & x_i \end{pmatrix} \left[\frac{1}{S_{xx}} \begin{pmatrix} \sum x_i^2/n & -\bar{x} \\ -\bar{x} & 1 \end{pmatrix} \right] \begin{pmatrix} 1 \\ x_i \end{pmatrix}$$

$$= \left[\frac{1}{S_{xx}} \right] \left[\frac{\sum x_i^2}{n} - x_i\bar{x} - x_i\bar{x} + x_i^2 \right]$$

$$= \left[\frac{1}{S_{xx}} \right] \left[\frac{\sum x_i^2}{n} - (\bar{x}^2) + (\bar{x}^2) - 2x_i\bar{x} + x_i^2 \right]$$

$$= \left[\frac{1}{S_{xx}} \right] \left[\frac{\sum x_i^2 - n\bar{x}^2}{n} + (x_i - \bar{x})^2 \right]$$

$$= \frac{1}{n} + \frac{(x_i - \bar{x})^2}{S_{xx}}$$

and

$$h_{ij} = \begin{pmatrix} 1 & x_i \end{pmatrix} \left[\frac{1}{S_{xx}} \begin{pmatrix} \sum x_i^2/n & -\bar{x} \\ -\bar{x} & 1 \end{pmatrix} \right] \begin{pmatrix} 1 \\ x_j \end{pmatrix}$$

$$= \left[\frac{1}{S_{xx}} \right] \left[\frac{\sum x_i^2}{n} - x_i\bar{x} - x_j\bar{x} + x_i x_j \right]$$

$$= \left[\frac{1}{S_{xx}} \right] \left[\frac{\sum x_i^2}{n} - (\bar{x})^2 + (\bar{x}^2) - x_i\bar{x} - x_j\bar{x} + x_i x_j \right]$$

$$= \left[\frac{1}{S_{xx}} \right] \left[\frac{\sum x_i^2 - n\bar{x}^2}{n} + (x_i - \bar{x})(x_j - \bar{x}) \right]$$

$$= \frac{1}{n} + \frac{(x_i - \bar{x})(x_j - \bar{x})}{S_{xx}}$$

3.30

$$\widehat{\beta} = (\mathbf{X}'\mathbf{X})^{-1}\mathbf{X}'\mathbf{y}$$
$$= (\mathbf{X}'\mathbf{X})^{-1}\mathbf{X}'[\mathbf{X}\beta + \beta]$$
$$= (\mathbf{X}'\mathbf{X})^{-1}\mathbf{X}'\mathbf{X}\beta + (\mathbf{X}'\mathbf{X})^{-1}\mathbf{X}'\beta$$
$$= \beta + \mathbf{R}\beta$$

3.31 From equation 3.15b, we get that $\beta = (\mathbf{I} - \mathbf{H})\mathbf{y}$. So substituting for $\mathbf{y}$, we get

$$(\mathbf{I} - \mathbf{H})(\mathbf{X}\beta + \beta) = \mathbf{X}\beta - \mathbf{X}(\mathbf{X}'\mathbf{X})^{-1}\mathbf{X}'\mathbf{X}\beta + (\mathbf{I} - \mathbf{H})\beta$$
$$= (\mathbf{I} - \mathbf{H})\beta$$

3.32

$$SS_R(\beta) = \widehat{\beta}'\mathbf{X}'\mathbf{y}$$
$$= \mathbf{y}'\mathbf{X}(\mathbf{X}'\mathbf{X})^{-1}\mathbf{X}'\mathbf{y}$$
$$= \mathbf{y}'\mathbf{H}\mathbf{y}$$

3.33

$$[\text{Corr}(\mathbf{y}, \widehat{\mathbf{y}})]^2 = \frac{[\mathbf{y}'\widehat{\mathbf{y}}]^2}{(\mathbf{y}'\mathbf{y})(\widehat{\mathbf{y}}'\widehat{\mathbf{y}})}$$
$$= \frac{[\mathbf{y}'\mathbf{H}\mathbf{y}]^2}{(\mathbf{y}'\mathbf{y})(\mathbf{y}'\mathbf{H}\mathbf{y})}$$
$$= \frac{(SS_R)^2}{(S_{yy})(SS_R)}$$
$$= SS_R/S_{yy} = R^2$$

3.34 $S = (\mathbf{y} - \mathbf{X}\beta)^{-1}(\mathbf{y} - \mathbf{X}\beta) - 2\lambda(\mathbf{T}\beta - \mathbf{c})$. Then take the derivative of S with respect to β and λ and set them equal to zero.

$$\frac{\partial S}{\beta} = -2\mathbf{X}'\mathbf{y} + 2(\mathbf{X}'\mathbf{X})^{-1}\widetilde{\beta} - 2\lambda\mathbf{T}' = \mathbf{0}, \frac{\partial S}{\lambda} = 2(\mathbf{T}\widetilde{\beta} - \mathbf{c}) = \mathbf{0}$$

This yields $\widetilde{\beta} = (\mathbf{X}'\mathbf{X})^{-1}\mathbf{X}'\mathbf{y} - (\mathbf{X}'\mathbf{X})^{-1}\mathbf{T}'\lambda$. Now substitute this expression for $\widetilde{\beta}$ into $\dfrac{\partial S}{\lambda}$ and solve for λ.

$$\mathbf{T}[(\mathbf{X}'\mathbf{X})^{-1}\mathbf{X}'\mathbf{y} - (\mathbf{X}'\mathbf{X})^{-1}\mathbf{T}'\lambda] - \mathbf{c} = \mathbf{0}$$
$$\mathbf{T}(\mathbf{X}'\mathbf{X})^{-1}\mathbf{T}' = \mathbf{T}\widehat{\beta} - \mathbf{c}$$
$$\lambda = [\mathbf{T}(\mathbf{X}'\mathbf{X})^{-1}\mathbf{T}']^{-1}(\mathbf{T}\widehat{\beta} - \mathbf{c})$$

Finally, substitute λ back into the equation for $\tilde{\boldsymbol{\beta}}$ which gives the desired result. Note that the sign will change when you write the last part as $\mathbf{c} - \mathbf{T}\widehat{\boldsymbol{\beta}}$.

3.35 The variance of $\widehat{\beta}_j$ is the jth diagonal element of $\sigma^2(\mathbf{X}'\mathbf{X})^{-1}$. Let $\mathbf{x}_j$ be the column of $\mathbf{X}$ associated with the jth regressor, and let $\mathbf{X}_{-j}$ be the rest of $\mathbf{X}$. Therefore,

$$\sigma^2(\mathbf{X}'\mathbf{X})^{-1} = \sigma^2 \begin{bmatrix} \mathbf{X}'_{-j}\mathbf{X}_{-j} & \mathbf{X}'_{-j}\mathbf{x}_j \\ \mathbf{x}'_j\mathbf{X}_{-j} & \mathbf{x}'_j\mathbf{x}_j \end{bmatrix}.$$

From Appendix C.2.1.13, the jth diagonal element of $\sigma^2(\mathbf{X}'\mathbf{X})^{-1}$ is

$$\mathbf{var}(\widehat{\beta}_j) = \sigma^2[\mathbf{x}'_j[\mathbf{I} - \mathbf{X}_{-j}(\mathbf{X}'_{-j}\mathbf{X}_{-j})^{-1}\mathbf{X}'_{-j}]\mathbf{x}_j]^{-1}$$

$$= \sigma^2[\mathbf{x}'_j\mathbf{x}_j - \mathbf{x}'_j\mathbf{X}_{-j}(\mathbf{X}'_{-j}\mathbf{X}_{-j})^{-1}\mathbf{X}'_{-j}\mathbf{x}_j]^{-1}$$

3.36 Since $R^2 = SS_R/S_{yy}$, we need to show that the sum of squares for regression for model B, SS_{Rb} is greater than the sum of squares for regression for model A, SS_{Ra}. We can do this using partitioning SS_R into sequential sums of squares. Consider i parameters in $\boldsymbol{\beta}_1$ and j parameters in $\boldsymbol{\beta}_2$. Then model B is using $(i \times j)$ parameters of which the first i are the same as model A. Then SS_{Rb} equals

$$R(\beta_{i1}, \beta_{i2}, \dots, \beta_{ii}, \beta_{j1}, \beta_{j2}, \dots, \beta_{jj}|\beta_0) = R(\beta_{i1}, \beta_{i2}, \dots, \beta_{ii}|\beta_0)$$

$$+ R(\beta_{j1}, \beta_{j2}, \dots, \beta_{jj}|\beta_0, \beta_{i1}, \beta_{i2}, \dots, \beta_{ii})$$

Since the second term on the right is a sum of squares, it must be greater than or equal to zero. Thus, $SS_{Rb} \geq SS_{Ra}$ which is equivalent to $R_B^2 \geq R_A^2$.

3.37 $\widehat{\boldsymbol{\beta}}_1 = (\mathbf{X}'_1\mathbf{X}_1)^{-1}\mathbf{X}_1\mathbf{y}$. Therefore,

$$E(\widehat{\boldsymbol{\beta}}_1) = (\mathbf{X}'_1\mathbf{X}_1)^{-1}\mathbf{X}'_1 E(\mathbf{y})$$

$$= (\mathbf{X}'_1\mathbf{X}_1)^{-1}\mathbf{X}'_1(\mathbf{X}_1\boldsymbol{\beta}_1 + \mathbf{X}_2\boldsymbol{\beta}_2)$$

$$= (\mathbf{X}'_1\mathbf{X}_1)^{-1}\mathbf{X}'_1\mathbf{X}_1\boldsymbol{\beta} + (\mathbf{X}'_1\mathbf{X}_1)^{-1}\mathbf{X}'_1\mathbf{X}_2\boldsymbol{\beta}_2$$

$$= \boldsymbol{\beta}_1 + (\mathbf{X}'_1\mathbf{X}_1)^{-1}\mathbf{X}'_1\mathbf{X}_2\boldsymbol{\beta}_2$$

The estimate is unbiased if $\mathbf{X}'_1\mathbf{X}_2$ is $\mathbf{0}$, which happens if $\mathbf{X}_1$ and $\mathbf{X}_2$ are orthogonal.

3.38

$$\sum_{i-1}^{n} Var(\widehat{y}_i) = \sum_{i=1}^{n} \mathbf{x}_i'(\mathbf{X'X})^{-1}\mathbf{x}_i(\sigma^2)$$

$$= \sigma^2 \left(\sum_{i=1}^{n} h_{ii} \right)$$

$$= \sigma^2(rank\,\mathrm{of}\,\mathbf{X}) = p\sigma^2$$

3.39 The Jth VIF is the jth diagonal element of $(\mathbf{W'W})^{-1}$, where $\mathbf{W'W}$ is the correlation matrix. Let $\mathbf{w}_j$ be the column of $\mathbf{W}$ associated with the jth regressor, and let $\mathbf{W}_{-j}$ be the rest of $\mathbf{W}$. Therefore,

$$(\mathbf{W'W})^{-1} = \begin{bmatrix} \mathbf{W}_{-j}'\mathbf{W}_{-j} & \mathbf{W}_{-j}'\mathbf{w}_j \\ \mathbf{w}_j'\mathbf{W}_{-j} & \mathbf{w}_j'\mathbf{w}_j \end{bmatrix}.$$

We note that $\mathbf{W'W}$ is the correlation matrix. As a result $\mathbf{w}_j'\mathbf{w}_j = 1$. From Appendix C.2.1.13, the jth diagonal element of $(\mathbf{W'W})^{-1}$ is

$$[\mathbf{w}_j'[\mathbf{I} - \mathbf{W}_{-j}(\mathbf{W}_{-j}'\mathbf{W}_{-j})^{-1}\mathbf{W}_{-j}']\mathbf{w}_j]^{-1}$$
$$= [\mathbf{w}_j'\mathbf{w}_j - \mathbf{w}_j'\mathbf{W}_{-j}(\mathbf{W}_{-j}'\mathbf{W}_{-j})^{-1}\mathbf{W}_{-j}'\mathbf{w}_j]^{-1}$$
$$= [1 - \mathbf{w}_j'\mathbf{W}_{-j}(\mathbf{W}_{-j}'\mathbf{W}_{-j})^{-1}\mathbf{W}_{-j}'\mathbf{w}_j]^{-1}.$$

Since $\mathbf{1'w}_j = 0$. If we regress $\mathbf{w}$ on $\mathbf{W}_{-j}$, we obtain that

$$\mathbf{SS}_{total} = \mathbf{w}_j'\mathbf{w}_j - \mathbf{w}_j'\mathbf{1}(\mathbf{1'1})^{-1}\mathbf{1'w}_j = 1,$$

and

$$\mathbf{SS}_{reg} = \mathbf{w}_j'\mathbf{W}_{-j}(\mathbf{W}_{-j}'\mathbf{W}_{-j})^{-1}\mathbf{W}_{-j}'\mathbf{w}_j.$$

As a result, if we regress $\mathbf{w}_j$ on $\mathbf{W}_{-j}$, the resulting R_j^2 is

$$R_j^2 = \frac{\mathbf{SS}_{reg}}{\mathbf{SS}_{total}}$$
$$= \mathbf{w}_j'\mathbf{W}_{-j}(\mathbf{W}_{-j}'\mathbf{W}_{-j})^{-1}\mathbf{W}_{-j}'\mathbf{w}_j.$$

As a result, the jth diagonal element of $(\mathbf{W'W})^{-1}$ is

$$[1 - \mathbf{w}_j'\mathbf{W}_{-j}(\mathbf{W}_{-j}'\mathbf{W}_{-j})^{-1}\mathbf{W}_{-j}'\mathbf{w}_j]^{-1} = [1 - R_j^2]^{-1} = \frac{1}{1 - R_j^2}.$$

3.40 If $\beta \sim N(\mathbf{0}, \sigma^2 \mathbf{I})$, then $\mathbf{T}\widehat{\beta} - \mathbf{c} \sim N(\mathbf{T}\beta - \mathbf{c}, \mathbf{T}(\mathbf{X}'\mathbf{X})^{-1}\mathbf{T}'(\sigma^2))$. Note that the rank $[\mathbf{T}(\mathbf{X}'\mathbf{X})^{-1}\,\mathbf{T}'] = \text{rank}[\mathbf{T}] = q$. First, we need to show that

$$Q/\sigma^2 = (\mathbf{T}\widehat{\beta} - \mathbf{c})'[\mathbf{T}(\mathbf{X}'\mathbf{X})^{-1}\mathbf{T}']^{-1}(\mathbf{T}\widehat{\beta} - \mathbf{c})/\sigma^2$$

is distributed as χ_q^2, under H_0. Since $\widehat{\beta} = (\mathbf{X}'\mathbf{X})^{-1}\mathbf{X}'\mathbf{y}$, then

$$Q = (\mathbf{T}(\mathbf{X}'\mathbf{X})^{-1}\mathbf{X}'\mathbf{y} - \mathbf{c})'(\mathbf{T}(\mathbf{X}'\mathbf{X})^{-1}\mathbf{T}')^{-1}(\mathbf{T}(\mathbf{X}'\mathbf{X})^{-1}\mathbf{X}'\mathbf{y} - \mathbf{c})$$

Now $\mathbf{T}(\mathbf{X}'\mathbf{X})^{-1}\,\mathbf{X}'\mathbf{y} - \mathbf{c} = \mathbf{T}\,(\mathbf{X}'\mathbf{X})^{-1}\,\mathbf{X}'\,[\mathbf{y} - \mathbf{X}\mathbf{T}'\,(\mathbf{T}\mathbf{T}')^{-1}\,\mathbf{c}]$. Hence,

$$Q = [\mathbf{y} - \mathbf{X}\mathbf{T}'(\mathbf{T}\mathbf{T}')^{-1}\mathbf{c}]'(\mathbf{X}(\mathbf{X}'\mathbf{X})^{-1}\mathbf{T}')(\mathbf{T}(\mathbf{X}'\mathbf{X})^{-1}\mathbf{T}')^{-1}(\mathbf{T}(\mathbf{X}'\mathbf{X})^{-1}\mathbf{X}')$$
$$[\mathbf{y} - \mathbf{X}\mathbf{T}'(\mathbf{T}\mathbf{T}')^{-1}\mathbf{c}]$$

Thus Q is expressed as a quadratic form in the vector $\mathbf{y} - \mathbf{X}\mathbf{T}'\,(\mathbf{T}\mathbf{T}')^{-1}$ $\mathbf{c}$. It is straight-forward to verify that the inner matrix of Q is idempotent. Also, since under H_0, $\mathbf{T}\beta = \mathbf{c}$, the noncentrality parameter λ is zero. Thus $Q/\sigma^2 \sim \chi_q^2$. Now we consider $SS_E = \mathbf{y}\,[\mathbf{I} - \mathbf{X}\,(\mathbf{X}'\mathbf{X})^{-1}\,\mathbf{X}']\,\mathbf{y}$. SS_E can also be written as a quadratic form in terms of the vector $\mathbf{y} - \mathbf{X}\mathbf{T}'\,(\mathbf{T}\mathbf{T}')^{-1}\,\mathbf{c}$:

$$SS_E = [\mathbf{y} - \mathbf{X}\mathbf{T}'(\mathbf{T}\mathbf{T}')^{-1}\mathbf{c}]'[\mathbf{I} - \mathbf{X}(\mathbf{X}'\mathbf{X})^{-1}\mathbf{X}'][\mathbf{y} - \mathbf{X}\mathbf{T}'(\mathbf{T}\mathbf{T}')^{-1}\mathbf{c}].$$

Since the matrix in this quadratic form is still $(\mathbf{I} - \mathbf{H})$, is it clear it is idempotent and $\lambda = 0$. Thus SS_E/σ^2 is distributed χ_{n-p}^2 Note that

$$[\mathbf{I} - \mathbf{X}(\mathbf{X}'\mathbf{X})^{-1}\mathbf{X}'](\mathbf{X}(\mathbf{X}'\mathbf{X})^{-1}\mathbf{T}')(\mathbf{T}(\mathbf{X}'\mathbf{X})^{-1}\mathbf{T}')^{-1}(\mathbf{T}(\mathbf{X}'\mathbf{X})^{-1}\mathbf{X}') = 0$$

Therefore, SS_E/σ^2 and Q/σ^2 are independently distributed as central chi-square variables under H_0. Hence, $F = \dfrac{Q/p}{MS_E} \overset{H_0}{\sim} F_{q,n-p}$.

Now under the alternative, $\mathbf{T}\beta \neq \mathbf{c}$. Therefore, we get $\lambda = (\mathbf{T}\beta - \mathbf{c})'(\mathbf{T}(\mathbf{X}'\mathbf{X})^{-1}\,\mathbf{T}')^{-1}\,(\mathbf{T}\beta - \mathbf{c})$. Hence, $Q/\sigma^2 \overset{H_1}{\sim} (\chi_q^2)^*$ which is a noncentral chi-square. Thus $F = \dfrac{Q/p}{MS_E} \overset{H_1}{\sim} F_{q,n-p}^*$ which is a noncentral F-distribution.

3.41 a. $F = 16.52$ with $p = < 0.0001$, which is significant.

b. ERA and errors are significant, while strikeouts and runs allowed per game are not significant at the $\alpha = 0.05$ level.

Parameter estimates

Term	Estimate	Std error	t ratio	Prob>|t|
Intercept	181.7296	33.41307	5.44	<.00001[*]
ERA	−49.74629	18.68691	−2.66	0.0134[*]
Errors	−0.318412	0.113428	−2.81	0.0095[*]
SO	−0.008188	0.01643	−0.50	0.6226
RA/G	33.065074	18.06051	1.83	0.0791

c. The most insignificant regressor (strikeouts) was removed first and the model was refit with the remaining regressors. The R^2_{Adj} for the original model with strikeouts was 0.68. After removing the strikeouts, the R^2_{Adj} has increased to 0.69. This would suggest strikeouts should be removed from the model. Next, runs allowed per game was removed. The new R^2_{Adj} value decreases to 0.665. This would suggest that runs allowed per game should be part of the model along with ERA and errors.

3.42 a. $F = 230.35$ with $p = <0.0001$, which is significant.

b.

Parameter estimates

Term	Lower 95%	Upper 95%
Intercept	434.39662	651.97224
Median price/sqft	5.2191981	6.4141182
Population	$8.5368e^{-6}$	0.0001561

c. The $R^2 = 0.91$. This means that 91% of the variability in home rental price is explained by this regression model. The model is doing an adequate job of explaining variability.

3.43 e. None of the above

3.44 c. 0.78

3.45 a. $\widehat{y} = 164.3 - 15.67(3.8) - 0.1885(78) = 90.05$

b. (86.46, 93.61)

c. $\widehat{y} = 158.8 - 18.62(3.8) = 88.04$

d. (84.82, 91.29)

e. The width of the interval for the model with only ERA is 6.47, while the width of the interval for the model that has both ERA and errors is 7.15. The interval is narrower for the simple linear regression (only ERA) indicating that adding errors to the model has not improved the model's estimation capability.

Chapter 4: Model Adequacy Checking

4.1 a. There does not seem to be a problem with the normality assumption.

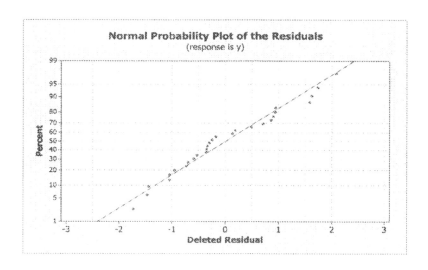

b. The model seems adequate.

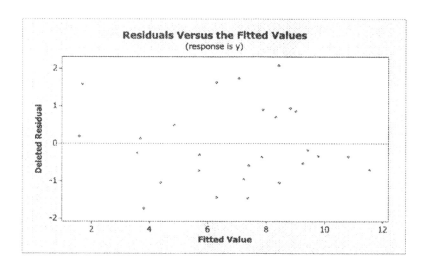

c. It appears that the model will be improved by adding x_2.

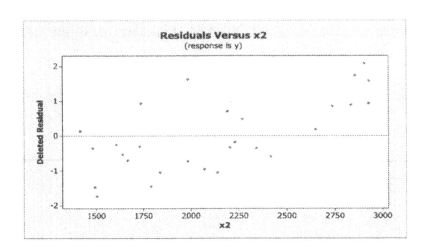

4.2   a. There looks to be a slight problem with normality.

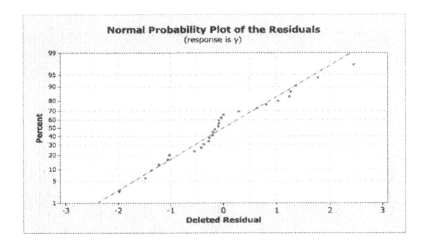

b. The plot looks good.

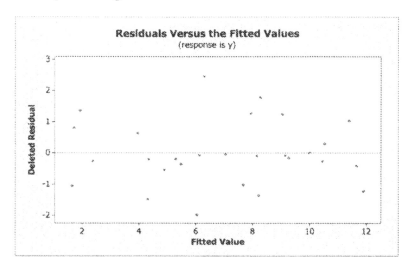

c. The plot for x_8 looks ok, the plot for x_2 shows mild nonconstant variance, and the plot for x_7 exhibits nonconstant variance.

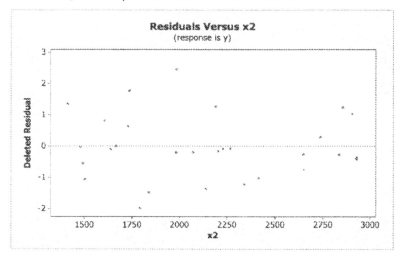

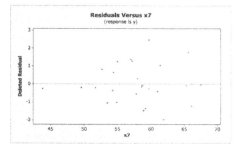

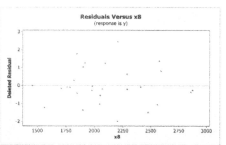

d. These plots indicate whether the relationships between the response and the regressor variables are correct. They show that there is not a strong linear relationship between the response and x_7.

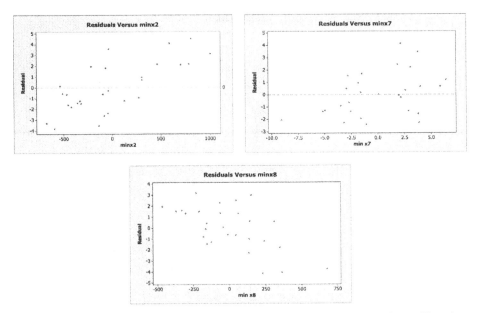

e. They can be used to determine influential points and outliers. For this example, the first observation is identified as a possible outlier.

4.3   a. There does not seem to be any problem with normality.

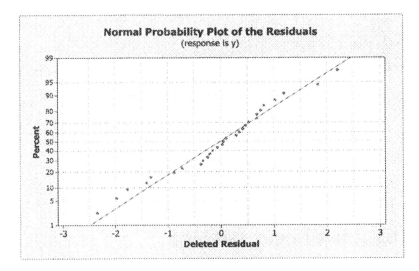

b. There appears to be a pattern and possible nonconstant variance.

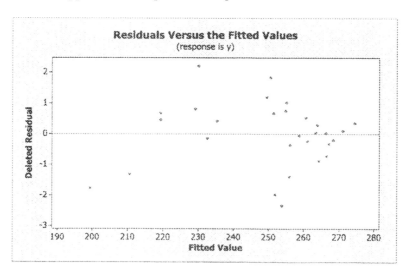

4.4 a. There seems to be a slight problem with normality.

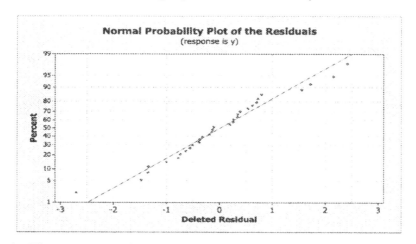

b. There appears to be a nonlinear pattern.

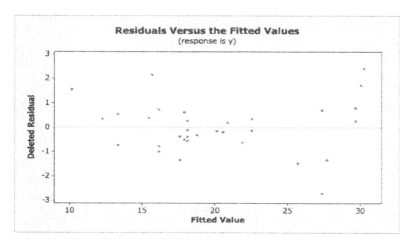

c. There is a linear pattern for x_1. The graph for x_6 shows no pattern and indicates it might be unnecessary to include it in the model.

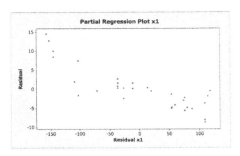
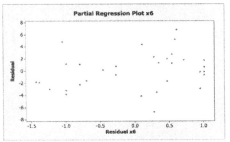

d. These residual indicate that observations 12 and 15 are possible outliers.

4.5 a. There does not appear to be a problem with normality.

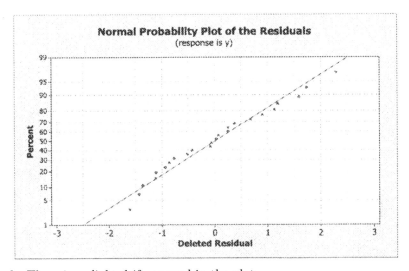

b. There is a slight drift upward in the plot.

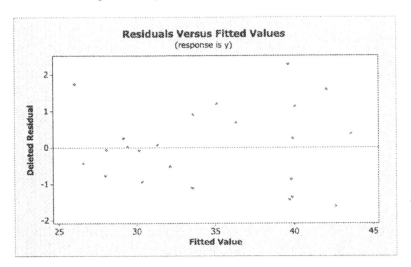

    c. Yes, after x_1 is in the model, most of the other variables contribute very
       little.

    d. They indicate observation 16 is a possible outlier.

4.6   a. There is a evidence of a problem with normality.

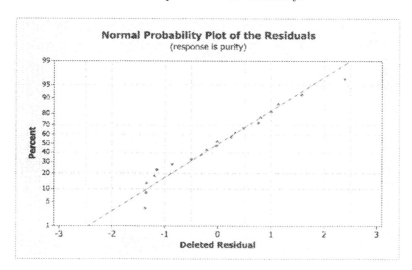

    b. There is a nonlinear pattern.

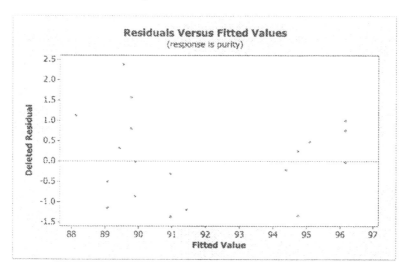

4.7   a. There is a serious problem with normality.

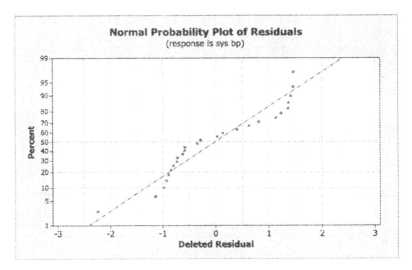

b. There is a nonlinear pattern.

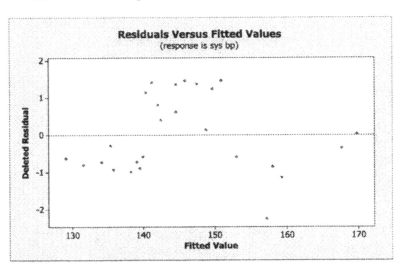

c. There does not appear to be any pattern with time.

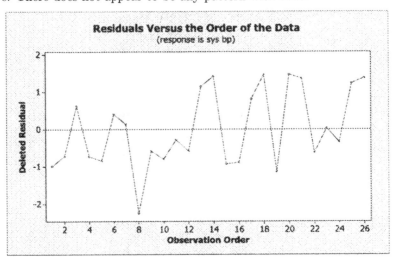

4.8   a. The plot shows normality is not a big problem.

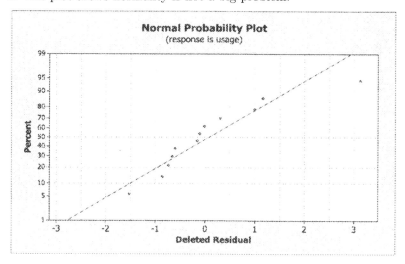

b. There is a pattern.

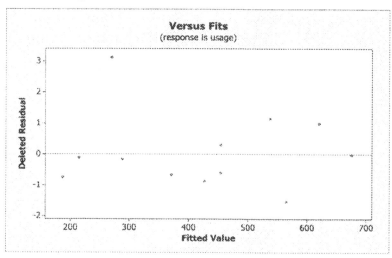

c. The plot shows positive autocorrelation.

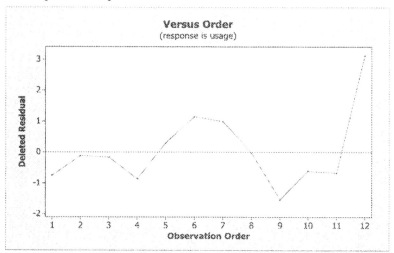

4.9   a. There appears to be no problem with normality.

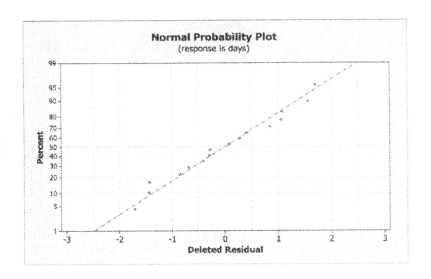

b. There is a pattern.

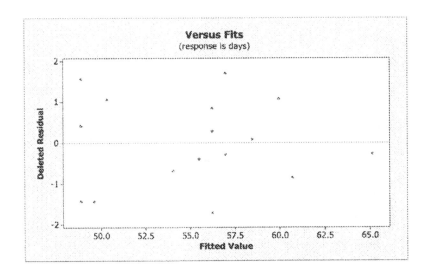

c. The plot shows positive autocorrelation.

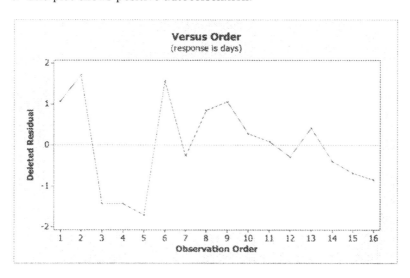

4.10 a. There appears to be no problem with normality.

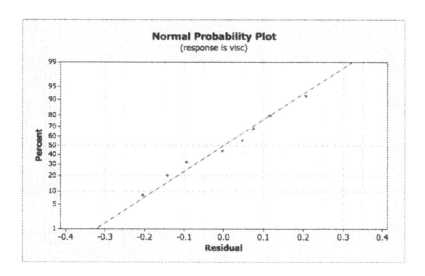

b. There is no real difference between the two plots.

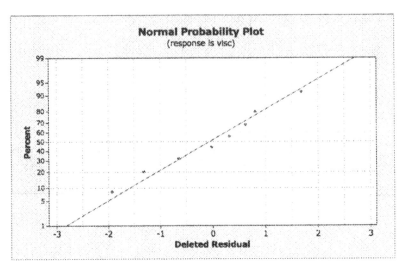

c. The plot shows a definite pattern.

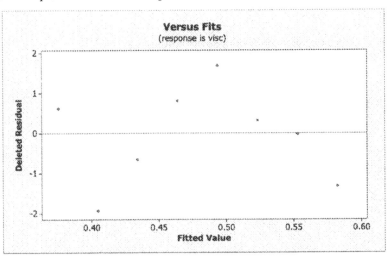

4.11 a. There is a slight problem with normality.

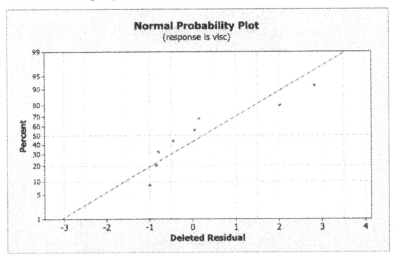

b. There is a quadratic pattern indicating that a second-order term is needed.

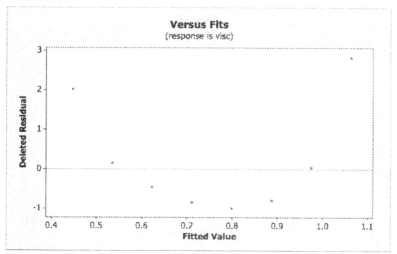

4.12 a. There is no problem with normality.

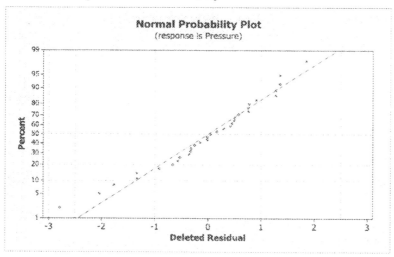

b. There is no pattern.

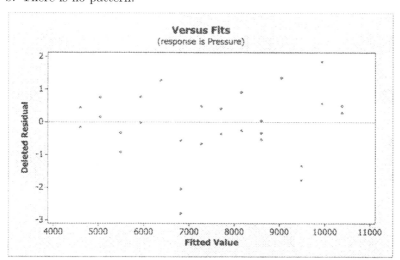

c. There is no pattern.

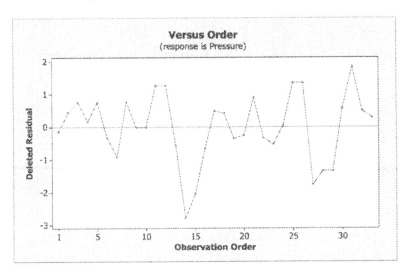

4.13 When x_7 and x_6 are in the model, PRESS $= 3388.6$ and $R^2_{Pred} = 56.94\%$. When just x_6 is in the model, PRESS $= 3692.9$ and $R^2_{Pred} = 53.08\%$. The residual plots for both models show nonconstant variance and departure from normality. There is no insight into the best choice of model.

4.14 When x_1 and x_6 are in the model, PRESS $= 328.8$ and $R^2_{Pred} = 73.43\%$. When just x_1 is in the model, PRESS $= 337.2$ and $R^2_{Pred} = 72.75\%$. Both models give basically the same values.

4.15 a. There does not seem to be a problem with normality.

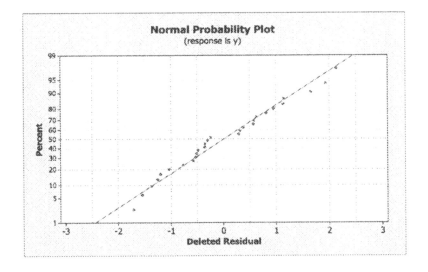

b. There is a nonlinear pattern.

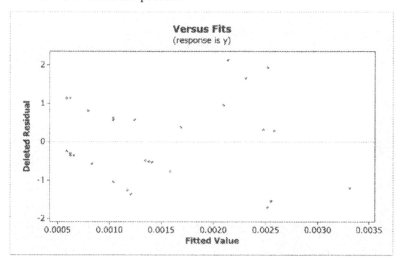

c. x_1 shows a linear pattern but x_4 does not.

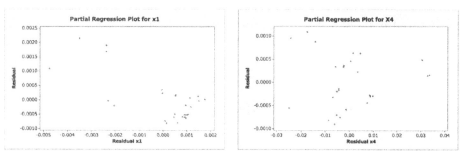

4.16 a. There is some problems in the tails.

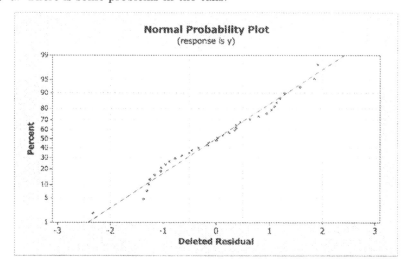

b. The fit seems pretty good.

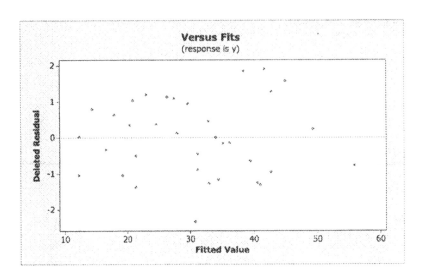

c. When x_1 and x_2 are in the model, PRESS = 916.41 and $R^2_{Pred} = 80.76\%$. When just x_2 is in the model, PRESS = 2825.62 and $R^2_{Pred} = 40.66\%$. The model with both x_1 and x_2 is more likely to provide better prediction of new data.

4.17 a. There is a serious problem with normality.

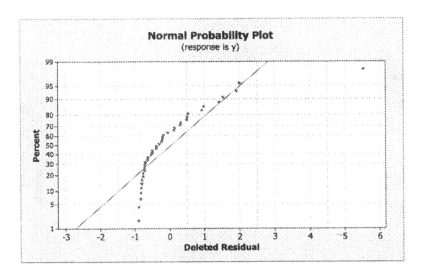

b. There is a nonlinear pattern.

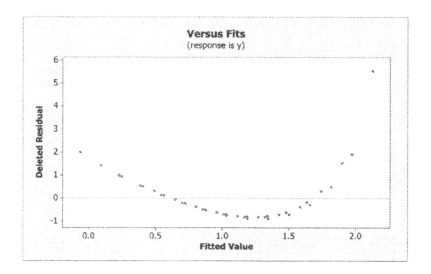

c. When x_1 and x_2 are in the model, PRESS = 3.11 and $R^2_{Pred} = 77.75\%$. When just x_2 is in the model, PRESS = 6.77 and $R^2_{Pred} = 51.54\%$. The model with both x_1 and x_2 is more likely to provide better prediction of new data.

4.18 a. Normality seems ok. There is a nonconstant variance problem. There is very little variability at the center points. The observation with $y = 55$ is a potential outlier.

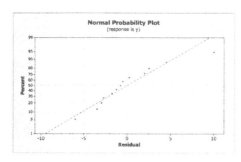
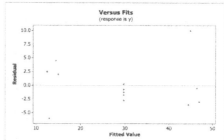

b. For lack of fit, $F_0 = \dfrac{39.16}{1.25} = 31.33$ with $p = 0.003$. There is evidence of lack of fit of the linear model.

4.19 a. Normality does not seem to be a problem. There is a nonlinear pattern in the residual plot versus the fitted values. The observation with $y = 198$ is a potential outlier.

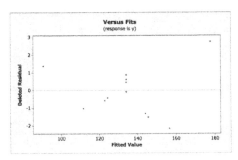 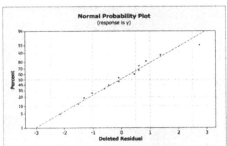

b. For lack of fit, $F_0 = \dfrac{299.5}{25.4} = 11.81$ with $p = 0.008$. There is evidence of lack of fit of the linear model.

4.20 a. There is a problem with normality. There is a problem with nonconstant variance. The observation with $y = 115.2$ is a potential outlier.

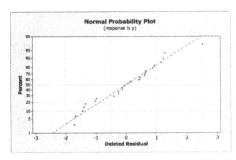 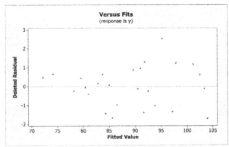

b. There is no test for lack of fit since there are no replicate points. It is possible to use the near-neighbor approach.

4.21 $E(MS_{PE}) = \dfrac{1}{n-m} E\left[\displaystyle\sum_{i=1}^{m}\sum_{j=1}^{n_i}(y_{ij}-\bar{y}_i)^2\right]$ where

$$E\left[\sum_{i=1}^{m}\sum_{j=1}^{n_i}(y_{ij}-\bar{y}_i)^2\right] = E\left[\sum_{i=1}^{m}\sum_{j=1}^{n_i}(y_{ij}^2 - 2y_{ij}\bar{y}_i + \bar{y}_i^2)\right]$$

$$= \sum_{i=1}^{m}\sum_{j=1}^{n_i}\left[E(y_{ij}^2) - 2E\left(y_{ij}\sum_{j^*=1}^{n_i}\frac{y_{ij^*}}{n_i}\right) + E(\bar{y}_i^2)\right]$$

$$= \sum_{i=1}^{m}\sum_{j=1}^{n_i}\left[\sigma^2 - 2E\left(y_{ij}\sum_{j^*=1}^{n_i}\frac{y_{ij^*}}{n_i}\right) + \frac{\sigma^2}{n_i}\right]$$

$$= n\sigma^2 - 2\sum_{i=1}^{m}\left(\sum_{j=1}^{n_i}\sum_{j^*=1}^{n_i}\frac{y_{ij}y_{ij^*}}{n_i}\right) + m\sigma^2$$

$$= n\sigma^2 - 2\sum_{i=1}^{m} \frac{n_i\sigma^2}{n_i} + m\sigma^2$$

$$= n\sigma^2 - 2m\sigma^2 + m\sigma^2$$

$$= (n - m)\sigma^2$$

Therefore, $E(MS_{PE}) = \sigma^2$.

Now, $SS_{Res} = SS_{PE} + SS_{LOF}$ and so $SS_{LOF} = SS_{RES} - SS_{PE}$. Using Appendix C for $E(SS_{Res})$ when the model is under specified and using $E(SS_{PE}) = (n - m)\sigma^2$ from above, we get

$$E(SS_{Res}) - E(SSPE) = (n - 2)\sigma^2 + \sum_{i=1}^{m} [E(y_i) - \beta_0 - \beta_1 x_i]^2 - (n - m)\sigma^2$$

$$= (m - 2)\sigma^2 + \sum_{i=1}^{m} [E(y_i) - \beta_0 - \beta_1 x_i]^2$$

Therefore,

$$E(MS_{LOF}) = E\left(\frac{SS_{LOF}}{m - 2}\right)$$

$$= \sigma^2 + \frac{\sum\limits_{i=1}^{m} [E(y_i) - \beta_0 - \beta_1 x_i]^2}{m - 2}$$

4.22 a. There is a problem with normality and there is a nonlinear pattern to the residual plot. Observation 2 is a potential outlier. The model does not fit well.

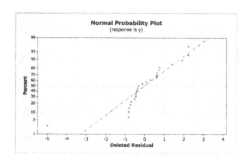
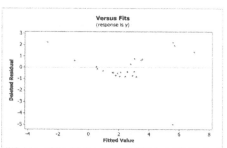

b. There is still a problem with normality and there is still a nonlinear pattern to the residual plot. Several observations are potential outliers. The model still does not fit well.

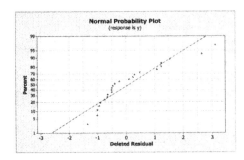 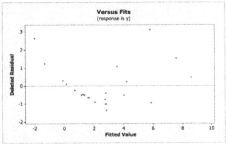

4.23 a. There does not appear to be a problem with normality.

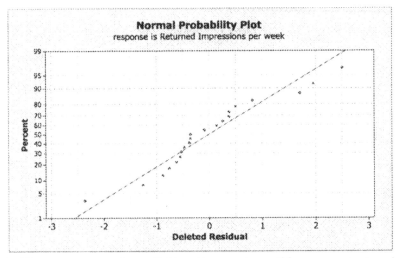

b. There appears to be a slight pattern and possible nonconstant variance.

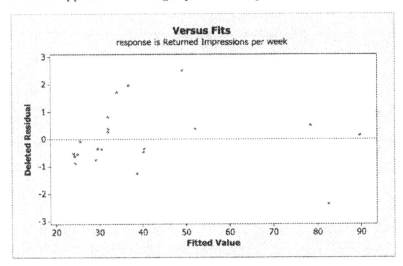

4.24 a. There does not appear to be a problem with normality.

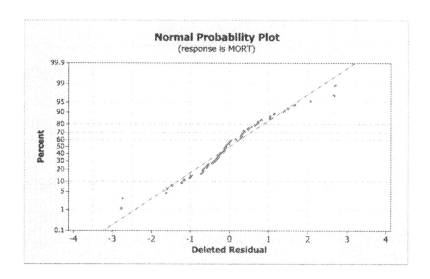

b. There is no apparent pattern.

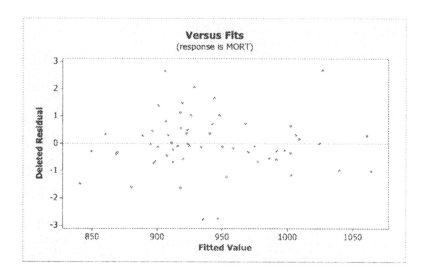

4.25 a. The plots for LifeExp and LifeExpMale show problems in the tails, but
      the LifeExpFemale plot shows no problems in normality.

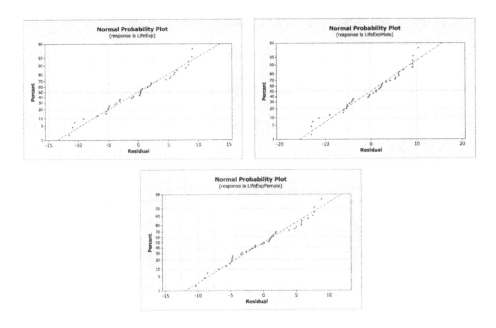

b. All three plots show a nonlinear pattern.

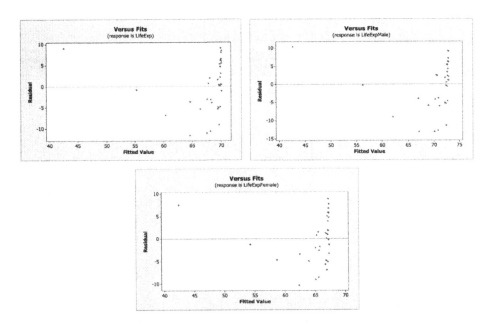

4.26 The normal probability plot indicates some possible deviations from nor-
      mality in the tails of the distribution; however this may be a result of
      observations 9 and 17 being possible outliers. The Deleted Residual versus

Fit plot also indicates that observations 9 and 17 are possible outliers but otherwise there is no apparent pattern.

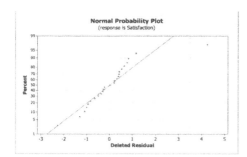
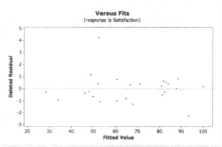

4.27 The residual analysis for the fuel consumption data indicates separation which may be a result of a variable missing from the model. There is also a pattern in the Deleted Residual versus Fit plot indicating the model is not adequate.

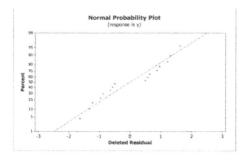
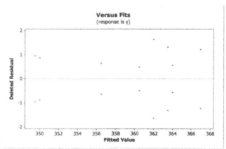

4.28 The residual analysis for the wine quality of young red wines data indicates an adequate model. There appears to be no problem with normality based on the normal probability plot and there is also no apparent pattern in the Deleted Residuals versus Fit plot.

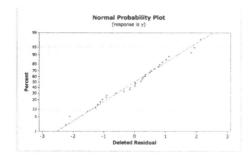
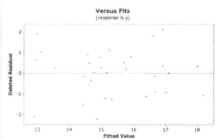

4.29 The residual analysis for the methanol oxidation data indicates no major problems with normality from the normal probability plot. However, the

Deleted Residuals versus Fits plot shows a nonlinear pattern indicating the model does not fit the data well.

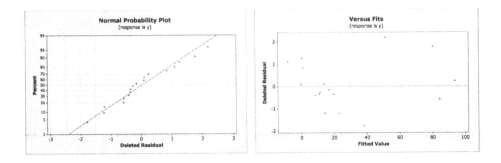

4.30 a. There does not seem to be a problem with the normality assumption. However, it may be useful to explore observation 28 (Texas Rangers) because it looks somewhat unusual.

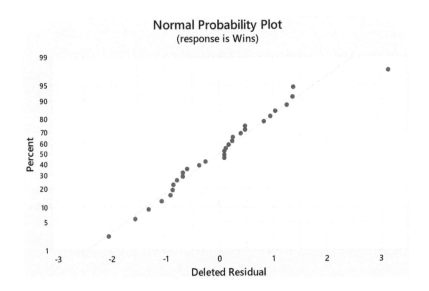

b. There does not seem to be any problem of model inadequacy.

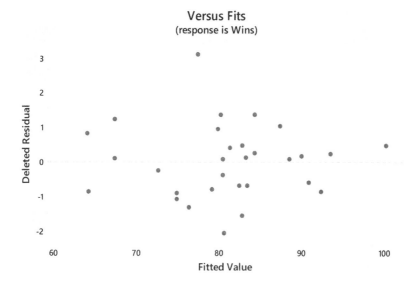

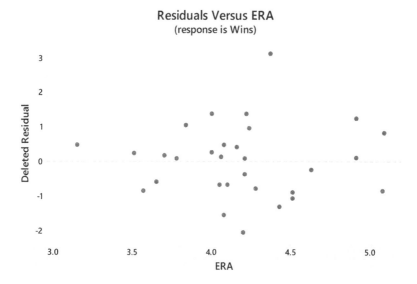

4.31 a. There does not seem to be a problem with the normality assumption. However, it may be useful to explore observation 28 (Texas Rangers) because it looks somewhat unusual.

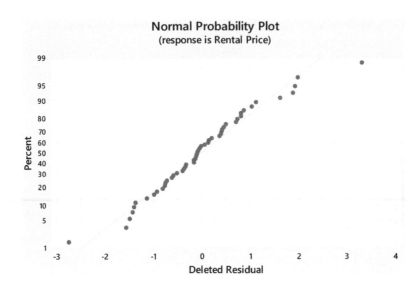

b. There does not seem to be a problem with model adequacy. Again, it may be useful to explore observation 28 (Texas Rangers) because of the large deleted residual.

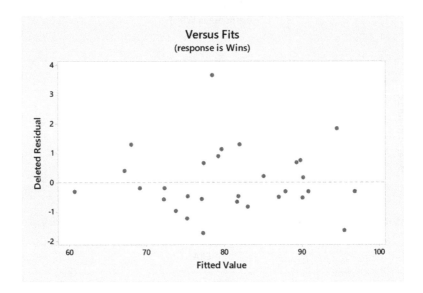

c. All plots look good. None present problems of model inadequacy.

4.32 a. There does not seem to be a problem with normality. It may be useful to explore observation 9 (Boston, MA) because of the large deleted residual.

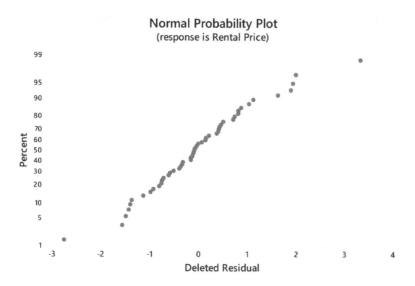

b. Both plots indicate some problems of model inadequacy with a potential problem of non-constant variance.

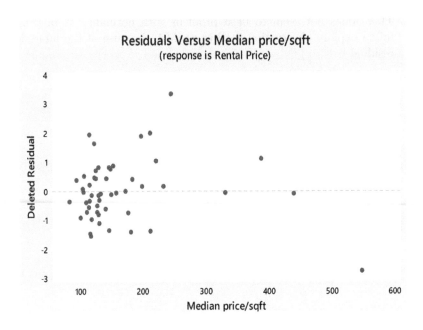

4.33 a. There does not seem to be a problem with normality. It may be useful to
          explore observation 9 (Boston, MA) because of the large deleted residual.

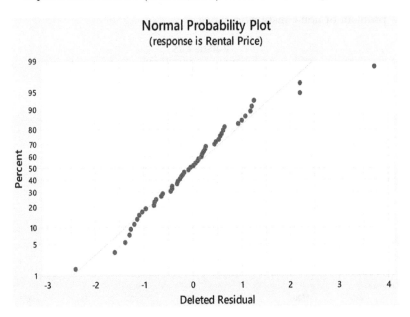

b. There seems to be an indication of possible non-constant variance.

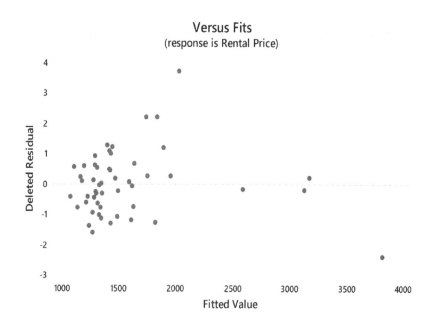

c. Both plots show some concern with constant variance.

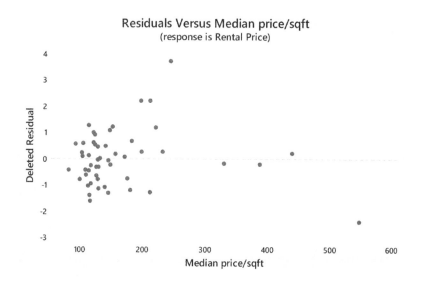

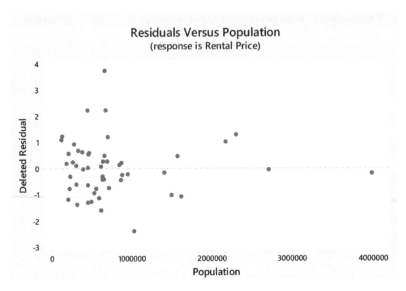

4.34 $PRESS = 1349.59$ and $R^2_{pred} = 59.11\%$

The model predicts new data at a reasonable level, but it could be improved.

4.35 a. $PRESS = 1314.70$ and $R^2_{pred} = 60.17\%$

b. The model predicts new data at a reasonable level, but it could be improved.

c. The PRESS for the simple linear regression was 1349.59 and for the multiple linear regression the value is 1314.70. Both models give basically the same values. This indicates you have not improved the predictive power with the more complicated model.

4.36 a. $PRESS = 1812163$ and $R^2_{pred} = 88.64\%$.

b. The model can predict new data very well.

Chapter 5: Transformations and Weighting to Correct Model Inadequacies

5.1   a. It has a nonlinear pattern.

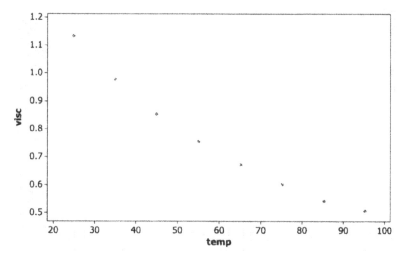

b. While $R^2 = 96\%$, the residual plot shows a nonlinear pattern and normality is violated.

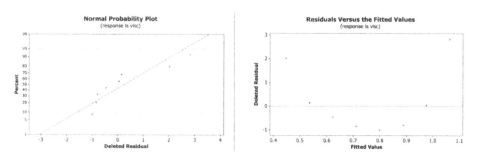

c. There is a slight improvement in the model.

5.2   a. There is a nonlinear pattern.

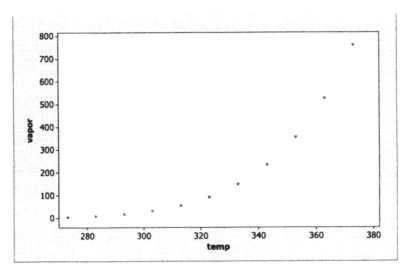

b. There is a problem with normality and a nonlinear pattern in the residuals.

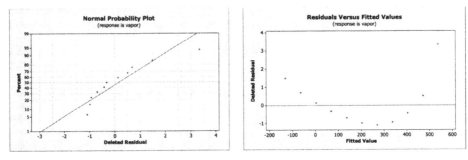

c. There is a slight improvement in the model.

5.3   a. There is a nonlinear pattern.

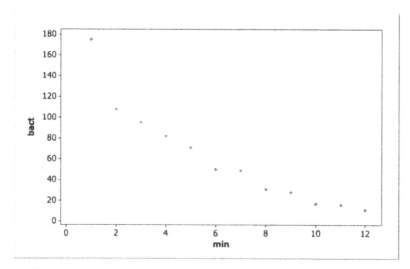

b. There is a problem with normality and a nonlinear pattern in the residuals. Observation 1 is an outlier.

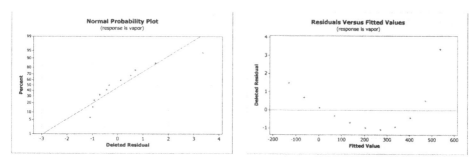

c. Fit the number of bacteria versus the natural log of the minutes. The first observation is still an outlier but otherwise the model fits fine.

5.4   The scatterplot looks fine. There is a problem with normality and the residual plot does not look good. Taking the natural log of x makes for a better model.

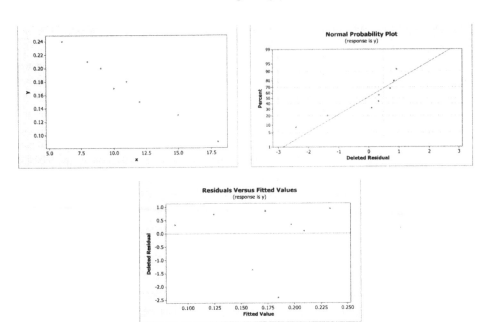

5.5   a. $\hat{y} = -31.698 + 7.277x$. There is a nonlinear pattern to the residuals.

       b. Taking the natural log of defects versus weeks makes for a better model.

5.6   a. The residual analysis from Exercise 4.27 indicates a problem with normality and a pattern in the residuals versus fits graph that indicates the model is not fitting the fuel consumption data well. However, this pattern does not suggest a transformation that would improve the analysis. Various transforms were applied but none improved the fit of the model. See problem 5.20 for an appropriate analysis of these data.

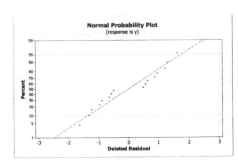

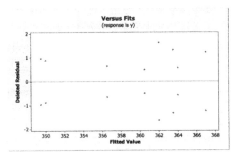

5.7   Prior to the residual analysis for the methanol oxidation data, the original
       model was reduced to only the significant regressors. This reduces the model
       from 5 regressors down to 2. This leaves regressors x_1 and x_3 in the model,
       reactor system and reactor residence time (seconds).
         The residual plots for this reduced model are seen below. There is a prob-
       lem with both the normality ansumption and there is also a pattern in the
       residual versus fits plot.

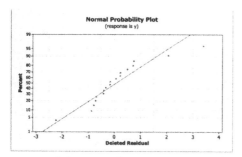
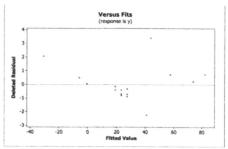

       A log transformation was performed on the response percent conversion.
       Regressor x_1 is not longer significant. The new regression equation is $\log(\hat{y}) =
       21.4 - 2.49x_3$. The estimate table is:

Coefficient	test statistic	p-value
β_3	-10.13	0.000

       The residuals plots below show no problem with the normality assumption
       and also show less of a pattern in the residual versus fits plot.

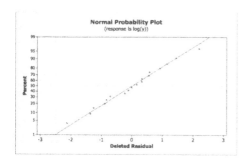
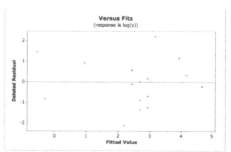

5.8   The models were sketched with $\beta_0 = 4$, $\beta_1 = 2$ and for $0 \leq x \leq 100$ by tens.
       The pattern is more consistent with a.

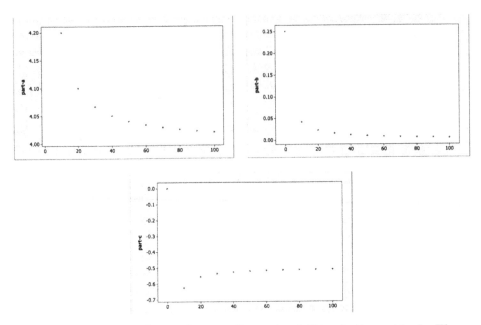

5.9   a. There is a problem with normality and a drifting in the residuals. There
           is an outlier at observation 28. x_2 has a nonlinear pattern.

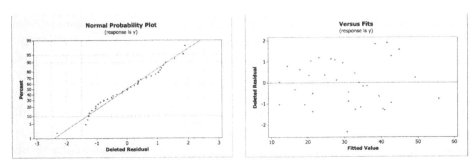

b. A square root transformation on y was used.

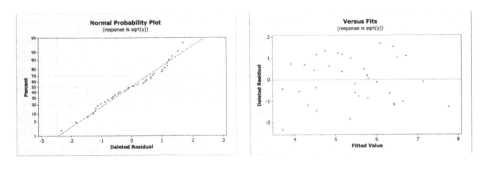

5.10 a. There is no problem with normality but a drifting in the residuals. There
are outliers. x_4 has a nonlinear pattern.

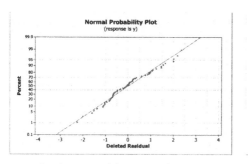

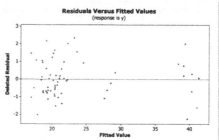

b. A natural log transformation on y was used.

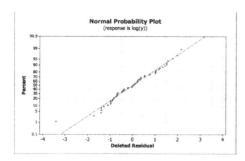

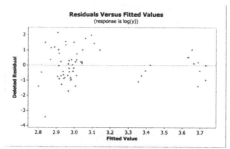

5.11 a. There is a problem with normality and a nonlinear pattern in the residuals. x_2 has a nonlinear pattern.

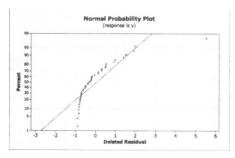

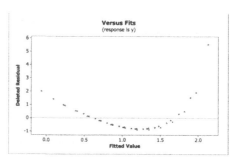

b. A transformation of $1/y$ was used along with inverting both of the independent variables.

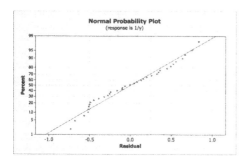

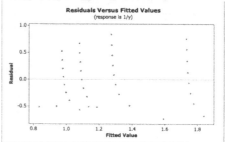

5.12 a. There is a departure from normality in the tail. There is a nonlinear pattern to the residuals. There is nonconstant variance. There are many potential outliers.

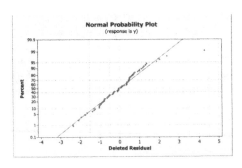 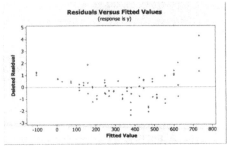

b. This corrects the nonconstant variance.

c. Use a square root transformation of the sample variance and model the sample standard deviation.

5.13 a. There is a departure from normality and a nonlinear pattern to the residuals.

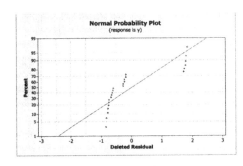 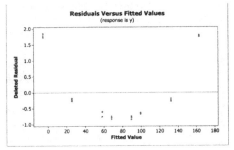

5.14 a. Yes, $\text{Var}(y_i') = \dfrac{1}{x_i^2}\text{Var}(y_i) = \sigma^2$.

b. Their roles are reversed.

c. The values of the parameters are the same but, by (b), their roles are reversed.

5.15 a. $S(\beta) = \sum\limits_{i=1}^{n} w_i(y_i - \beta x_i)^2$. Taking the derivative with respect to β and setting it equal to zero gives $\sum\limits_{i=1}^{n} w_i(Y_i - \hat{\beta}x_i)(-x_i) = 0$. Solving for $\hat{\beta}$ yields

$$\hat{\beta} = \frac{\sum\limits_{i=1}^{n} w_i x_i y_i}{\sum\limits_{i=1}^{n} w_i x_i^2}.$$

b.

$$\text{Var}(\hat{\beta}) = \left(\frac{1}{\displaystyle\sum_{i=1}^{n} w_i x_i^2} \right)^2 \sum_{i=1}^{n} w_i^2 x_i^2 \text{Var}(y_i)$$

$$= \left(\frac{1}{\displaystyle\sum_{i=1}^{n} w_i x_i^2} \right)^2 \sum_{i=1}^{n} w_i^2 x_i^2 (\sigma^2/w_i)$$

$$= \frac{\sigma^2}{\displaystyle\sum_{i=1}^{n} w_i x_i^2}$$

c. Here, we have $w_i = 1/x_i$. Therefore,

$$\hat{\beta} = \frac{\displaystyle\sum_{i=1}^{n} (1/x_i) x_i y_i}{\displaystyle\sum_{i=1}^{n} (1/x_i) x_i^2}$$

$$= \frac{\displaystyle\sum_{i=1}^{n} y_i}{\displaystyle\sum_{i=1}^{n} x_i}$$

with $\text{Var}(\hat{\beta}) = \dfrac{\sigma^2}{\displaystyle\sum_{i=1}^{n} x_i}$.

d. Here, we have $w_i = 1/x_i^2$. Therefore,

$$\hat{\beta} = \frac{\displaystyle\sum_{i=1}^{n} (1/x_i^2) x_i y_i}{\displaystyle\sum_{i=1}^{n} (1/x_i^2) x_i^2} = (1/n) \sum_{i=1}^{n} \frac{y_i}{x_i}$$

with $\text{Var}(\hat{\beta}) = \dfrac{\sigma^2}{n}$.

5.16 Let $\beta = \binom{\beta_1}{\beta_2}$, p_2 be the number of parameters in β_2, $\mathbf{K}' = (\mathbf{0}\ \mathbf{I})$, $\mathbf{m} = \mathbf{0}$, and the rank of $\mathbf{K}' = p_2$. Note this gives $\mathbf{K}'\hat{\beta} = \hat{\beta}_2$. Then the appropriate test statistic is

$$F_0 \frac{(\mathbf{K}'\hat{\beta} - \mathbf{m})'[\mathbf{K}'(\mathbf{X}'\mathbf{X})^{-1}\mathbf{K}]^{-1}(\mathbf{K}'\hat{\beta} - \mathbf{m})}{p_2 MSE}$$

Now under H_0, F_0 above has a central F distribution and under H_1 it has a noncentral F distribution.

5.17 Notice that we can write the top as the quadratic form,

$$\mathbf{y}'[\mathbf{V}^{-1} - \mathbf{V}^{-1}\mathbf{X}(\mathbf{X}'\mathbf{V}^{-1}\mathbf{X})^{-1}\mathbf{X}'\mathbf{V}^{-1}]\mathbf{y}.$$

Call the matrix in the brackets $\mathbf{A}$. Then from Appendix C, we get $E(\mathbf{y}'\mathbf{A}\mathbf{y})$ $= \text{trace}[(\mathbf{A})(\sigma^2\mathbf{V})] + \mu'\mathbf{A}\mu$ where for us, $\mu = E(\mathbf{y}) = \mathbf{0}$. It is easy to show that $[\mathbf{A}\mathbf{V}]$ is idempotent, so it's trace is equal to its rank, which is $n - p$. Thus, in this case, $E(\mathbf{y}'\mathbf{A}\mathbf{y}) = \text{trace}[(\mathbf{A})(\sigma^2\mathbf{V})] + \mu'\mathbf{A}\mu = (n - p)\sigma^2$.

5.18 a. There is a nonlinear pattern to the residuals.

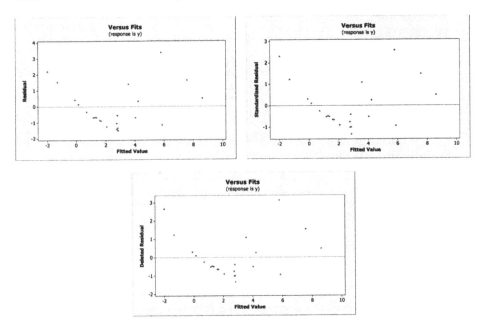

b. Use a natural log transformation on y. It does not improve the model.

c. Use a natural log transformation on each of the regressors in addition to the transformation in part b.

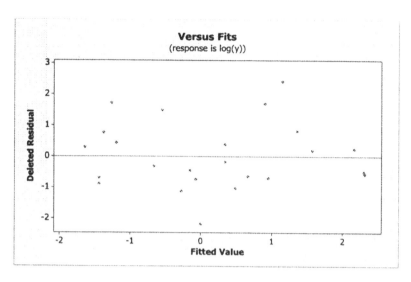

5.19 a.

$$Var(\mathbf{y}) = Var(\mathbf{X}\beta + \mathbf{Z}\delta + \epsilon)$$
$$= \mathbf{Z}Var(\delta)\mathbf{Z}' + Var(\epsilon)$$
$$= \mathbf{Z}(\sigma_\delta^2\mathbf{I}_m)\mathbf{Z}' + \sigma^2\mathbf{I}$$
$$= \sigma_\delta^2\mathbf{Z}\mathbf{Z}' + \sigma^2\mathbf{I}.$$

b. From part a, we have $Var(\mathbf{y}) = \sigma^2\mathbf{I} + \sigma_\delta^2\mathbf{Z}\mathbf{Z}' = \Sigma$.
Then

$$\mathbf{I} = \Sigma\Sigma^{-1}$$

$$= [\sigma^2\mathbf{I} + \sigma_\delta^2\mathbf{Z}\mathbf{Z}']\left[\frac{1}{\sigma^2}\mathbf{I} - k\mathbf{Z}\mathbf{Z}'\right].$$

In order to solve for Σ^{-1}, we must solve for k. Multiplying $\Sigma\Sigma^{-1}$ leads us to setting the following quantity equal to 0.

$$0 = -k\sigma^2\mathbf{Z}\mathbf{Z}' + \frac{\sigma_\delta^2}{\sigma^2}\mathbf{Z}\mathbf{Z}' - k\sigma_\delta^2\mathbf{Z}\mathbf{Z}'\mathbf{Z}\mathbf{Z}'$$

$$= \mathbf{Z}\left[-k\sigma^2\mathbf{I} + \frac{\sigma_\delta^2}{\sigma^2}\mathbf{I} - k\sigma_\delta^2\mathbf{Z}'\mathbf{Z}\right]\mathbf{Z}'.$$

Therefore,

$$0 = -k\sigma^2\mathbf{I} + \frac{\sigma_\delta^2}{\sigma^2}\mathbf{I} - k\sigma_\delta^2\mathbf{Z}'\mathbf{Z}$$

$$= -k\sigma^2\mathbf{I} + \frac{\sigma_\delta^2}{\sigma^2}\mathbf{I} - kn\sigma_\delta^2\mathbf{I}.$$

We solve for k

$$k = \frac{\sigma_\delta^2}{\sigma^2(\sigma^2 + n\sigma_\delta^2)}.$$

Then

$$\Sigma^{-1} = \frac{1}{\sigma^2}\mathbf{I} - \frac{\sigma_\delta^2}{\sigma^2(\sigma^2 + n\sigma_\delta^2)}\mathbf{Z}\mathbf{Z}'.$$

Now we must show $(\mathbf{X}'\Sigma^{-1}\mathbf{X})^{-1}\mathbf{X}'\Sigma^{-1}\mathbf{y} = (\mathbf{X}'\mathbf{X})^{-1}\mathbf{X}'\mathbf{y}$. First let's solve $\mathbf{X}'\Sigma^{-1}\mathbf{X}$

$$\mathbf{X}'\Sigma^{-1}\mathbf{X} = \mathbf{X}'[\frac{1}{\sigma^2}\mathbf{I} - \frac{\sigma_\delta^2}{\sigma^2(\sigma^2 + n\sigma_\delta^2)}\mathbf{Z}\mathbf{Z}']\mathbf{X}$$

$$= \frac{1}{\sigma^2}\mathbf{X}'\mathbf{X} - \frac{\sigma_\delta^2}{\sigma^2(\sigma^2 + n\sigma_\delta^2)}\mathbf{X}'\mathbf{Z}\mathbf{Z}'\mathbf{X}$$

$$= \frac{1}{\sigma^2}\mathbf{X}'\mathbf{X} - \frac{n\sigma_\delta^2}{\sigma^2(\sigma^2 + n\sigma_\delta^2)}\mathbf{X}'\mathbf{X}$$

$$= \frac{1}{\sigma^2 + n\sigma_\delta^2}\mathbf{X}'\mathbf{X}.$$

Now let's solve $\mathbf{X}'\Sigma^{-1}$

$$\mathbf{X}'\Sigma^{-1} = \mathbf{X}'\left[\frac{1}{\sigma^2}\mathbf{I} - \frac{\sigma_\delta^2}{\sigma^2(\sigma^2 + n\sigma_\delta^2)}\mathbf{Z}\mathbf{Z}'\right]$$

$$= \frac{1}{\sigma^2}\mathbf{X}' - \frac{\sigma_\delta^2}{\sigma^2(\sigma^2 + n\sigma_\delta^2)}\mathbf{X}'\mathbf{Z}\mathbf{Z}'$$

$$= \frac{1}{\sigma^2}\mathbf{X}' - \frac{n\sigma_\delta^2}{\sigma^2(\sigma^2 + n\sigma_\delta^2)}\mathbf{X}'$$

$$= \frac{1}{\sigma^2 + n\sigma_\delta^2}\mathbf{X}'.$$

As a result, $(\mathbf{X}'\Sigma^{-1}\mathbf{X})^{-1}\mathbf{X}'\Sigma^{-1}\mathbf{y}$ becomes

$$(\mathbf{X}'\Sigma^{-1}\mathbf{X})^{-1}\mathbf{X}'\Sigma^{-1}\mathbf{y} = (\sigma^2 + n\sigma_\delta^2)(\mathbf{X}'\mathbf{X})^{-1}\frac{1}{(\sigma^2 + n\sigma_\delta^2)}\mathbf{X}'\mathbf{y}$$

$$= (\mathbf{X}'\mathbf{X})^{-1}\mathbf{X}'\mathbf{y}.$$

This proves that the ordinary least squares estimates for β are the same as the generalized least squares estimates.

5.20 The proper analysis for the fuel consumption data is a regression analysis on the difference in fuel consumption (y) based on the batch. Because the batches of oil were divided into two with one batch going to the bus and the other batch going to the truck, a regression analysis on the difference to overcome the effect of batch. Also, for the regression analysis, we reduced the model until only significant regressors were present in the model. This leaves regressors x_4 and x_5 in the model, viscosity and initial boiling point.

The new regression equation is $\hat{y}_{Difference} = -106 - 13.0x_4 + 0.651x_5$.

The estimate table is:

Coefficient	test statistic	p-value
β_4	-3.09	0.027
β_5	8.99	0.000

The residual plots for this reduced model are seen below. The analysis on the difference in fuel consumption has alleviated the problems identified in problem 4.27.

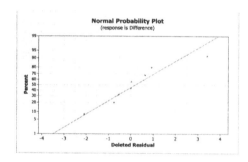
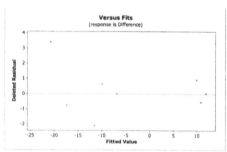

5.21 A regression analysis using 3 indicator variables for mix rate was carried out for the tensile strength data. (Note: An ANOVA anlaysis could be performed on this data. The results from the ANOVA are equivalent with the results from the regression analysis.)

The regression equation is $\hat{y} = 2666 + 430x_{150} + 490x_{175} + 267x_{200}$.

The estimate table is:

Coefficient	test statistic	p-value
β_{150}	5.83	0.000
β_{175}	6.65	0.000
β_{200}	3.63	0.003

The regression indicates that mix rate (rpm) has an effect on tensile strength. The p-values from the estimate table are computed from comparisons of average tensile strength from mix rates 150, 175, and 200 with mix rate 225. The average tensile strengths for mix rates 150, 175, and 200 are significantly higher compared to the tensile strength for a mix rate of 225.

The residual plots for this model seen below indicate no problems.

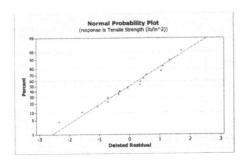 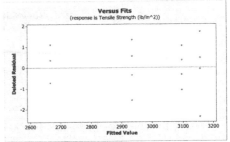

5.22 A regression analysis using 3 indicator variables for temperature was carried out for the density data. (Note: An ANOVA analysis could be performed on this data. The results from the ANOVA are equivalent with the results from the regression analysis.)

The regression equation is $\hat{y} = 23.2 - 1.46x_{900} - 0.700x_{910} - 0.280x_{920}$.

The estimate table is:

Coefficient	test statistic	p-value
β_{900}	-6.61	0.000
β_{910}	-3.00	0.009
β_{920}	-1.27	0.226

The regression indicates that peak kiln temperature has an effect on density of bricks. The p-values from the estimate table are computed from comparisons of average density from temperatures 900, 910, and 920 with temperature 930. The average density for temperatures 900 and 910 are significantly lower compared to the average density at 930. The average density is not significantly different for temperatures at 920 and 930.

The residual plots indicate a potential outlier in observation 10 (Temp. 920, Density 23.9).

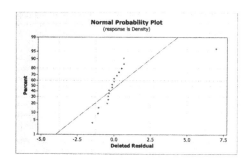
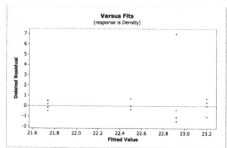

5.23 This Fixed Effects tests for the subsampling analysis indicates that the three vat pressures do not have a significant effect on strength ($F = 2.3984$, $p - value = 0.1716$). The variance component for batch is 0.743. A high percentage (73%) of the total variability is due to the batch-to-batch variability.

Effect Tests

Source	Nparm	DF	Sum of Squares	F Ratio	Prob > F
Pressure	2	2	4.2238889	2.3984	0.1716

REML Variance Component Estimates

Random Effect	Var Ratio	Var Component	Std Error	95% Lower	95% Upper	Pct of Total
Batch [Pressure]	2.7020202	0.7430556	0.5125044	−0.261435	1.7475457	72.988
Residual		0.275	0.1296362	0.1301072	0.9165344	27.012
Total		1.0180556				100.000

−2 LogLikelihood = $1.917470887

The plot of the residuals versus fits shows that the model is reasonable and the normal probability plot does not show a problem with the normality assumption.

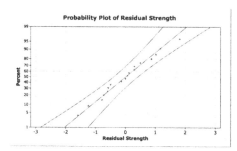
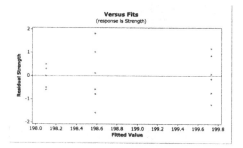

Chapter 6: Diagnostics for Leverage and Influence

6.1 Observation 1 is identified as influential. It affects the coefficients for x_3 and x_5.

6.2 No observations show up as influential.

6.3 Observation 14 is identified as influential. It seriously affects the coefficients for x_5 and x_6.

6.4 No observations show up as influential.

6.5 No observations show up as influential.

6.6 No observations show up as influential.

6.7 No observations show up as influential.

6.8 Observations 50-53 show up as influential.

6.9 Observation 31 shows up as influential.

6.10 Appendix C establishes that $\hat{\beta}_{(i)} - \hat{\beta} = \dfrac{(\mathbf{X'X})^{-1}\mathbf{x}_i e_i}{1 - h_{ii}}$. Therefore,

$$
\begin{aligned}
D_i &= \frac{\left(\hat{\beta}_{(i)} - \hat{\beta}\right)' \mathbf{X'X} \left(\hat{\beta}_{(i)} - \hat{\beta}\right)}{p MS_{Res}} \\
&= \frac{\mathbf{x}_i (\mathbf{X'X})^{-1} \mathbf{X'X} (\mathbf{X'X})^{-1} \mathbf{x}_i e_i^2}{(1 - h_{ii})^2 p MS_{Res}} \\
&= \left(\frac{e_i}{1 - h_{ii}}\right)^2 \left(\frac{h_{ii}}{p MS_{Res}}\right) \\
&= \left(\frac{e_i^2}{MS_{Res}(1 - h_{ii})}\right) \left(\frac{1}{p}\right) \left(\frac{h_{ii}}{1 - h_{ii}}\right) \\
&= \frac{r_i^2}{p} \left(\frac{h_{ii}}{1 - h_{ii}}\right)
\end{aligned}
$$

6.11 Appendix C establishes

$$
[\mathbf{X}_{(i)}' \mathbf{X}_{(i)}]^{-1} = (\mathbf{X'X})^{-1} + \frac{(\mathbf{X'X})^{-1} \mathbf{x}_i \mathbf{x}_i' (\mathbf{X'X})^{-1}}{1 - h_{ii}}
$$

Therefore,

$$COV\ RATIO_i = \frac{\left|(\mathbf{X}'_{(i)}\mathbf{X}_{(i)})^{-1}S^2_{(i)}\right|}{\left|(\mathbf{X}'\mathbf{X})^{-1}MS_{Res}\right|}$$

$$= \frac{\left(S^2_{(i)}\right)^p}{MS^p_{Res}} \frac{\left|(\mathbf{X}'\mathbf{X})^{-1} + \dfrac{(\mathbf{X}'\mathbf{X})^{-1}\mathbf{x}_i\mathbf{x}'_i(\mathbf{X}'\mathbf{X})^{-1}}{1-h_{ii}}\right|}{\left|(\mathbf{X}'\mathbf{X})^{-1}\right|}$$

$$= \left[\frac{S^2_{(i)}}{MS_{Res}}\right]^p \left(\frac{\mathbf{x}'_i(\mathbf{X}'\mathbf{X})^{-1}\mathbf{x}_i + \dfrac{\mathbf{x}'_i(\mathbf{X}'\mathbf{X})^{-1}\mathbf{x}_i\mathbf{x}'_i(\mathbf{X}'\mathbf{X})^{-1}\mathbf{x}_i}{1-h_{ii}}}{\mathbf{x}'_i(\mathbf{X}'\mathbf{X})^{-1}\mathbf{x}_i}\right)$$

(note the determinants have been dropped because they are scalars)

$$= \left[\frac{S^2_{(i)}}{MS_{Res}}\right]^p \left(\frac{h_{ii} + \dfrac{h^2_{ii}}{1-h_{ii}}}{h_{ii}}\right)$$

$$= \left[\frac{S^2_{(i)}}{MS_{Res}}\right]^p \left(\frac{1}{1-h_{ii}}\right)$$

6.12 No observations show up as influential.

6.13 The last observation shows up as influential.

6.14 Observation 20 shows up as influential.

6.15 Observations 2 and 4 show up as influential.

6.16 In looking at the plots of the residuals vs. the predictors, we can see a pattern with SO2.

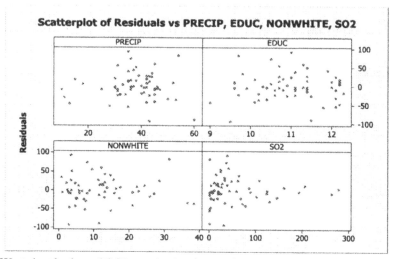

Scatterplot of Residuals vs PRECIP, EDUC, NONWHITE, SO2

We take the log of SO2 to obtain the model

$\widehat{y} = 942 - 13.8EDUC + 3.34NONWHITE + 1.67PRECIP + 34.3logSO2$
(Recall that NOX was not significant in our previous analyses.) The model
is significant with $F = 30.14$ and $p = 0.000$ with an $R^2 = 68.7\%$ and
$R^2_{Adj} = 66.4\%$. The residuals look fine plotted against the fitted values and
the individual regressors. None of the observations are influential.

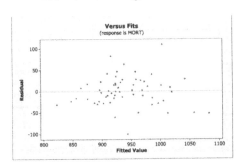
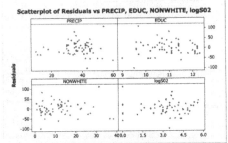

6.17 For all three models, we transform the data using square roots of both the
response and the regressors. For Life Expectancy, this gives the model,
$\widehat{y}^* = 8.67 - 0.0323sqrt(x_1) - 0.00713sqrt(x_2)$. $F = 30.25$ with $p = 0.000$,
so the model is significant. $R^2 = 63.4\%$ and $R^2_{Adj} = 61.3\%$. The residuals
look fine, except for the outlier from observation 8. Observations 8, 21, and
30 are influential for each model.

6.18 The regression analysis for the patient satisfaction data can be found in
section 3.6 of the text and the residual analysis can be found in Exercise 4.26.
The influence analysis for this regression indicates that observations 9 and
17 are highly influential.

6.19 From Exercise 5.20 we recognized that the analysis for the fuel consumption data requires an analysis on the difference in fuel consumption for
buses versus trucks. See Exercise 5.20 for the regression analysis of these
data. The residuals indicate observation 5 as a possible outlier. The influence analysis for this regression indicates that observation 5 is influential for
the model.

6.20 The regression analysis for the wine quality of young red wines data can be
found in Exercise 3.19 and the residual analysis can be found in Exercise 4.28.
The influence analysis for this regression indicates that observations 28 and
32 are highly influential.

6.21 The regression and residual analysis for the methanol oxidation data can be
found in Exercise 5.7. To improve the model we took a log transformation of
the response and reduced the model to only contain the significant predictor x_3. The influence analysis for this regression indicates that observation 1
is highly influential.

6.22 Cooks D detects highly influential points. The Cooks D value for observation
is higher compared with the rest of the data. This data point could be
influential. There are a couple of observations flagged as leverage points in
the Hats vs. Observation plot that could also be explored.

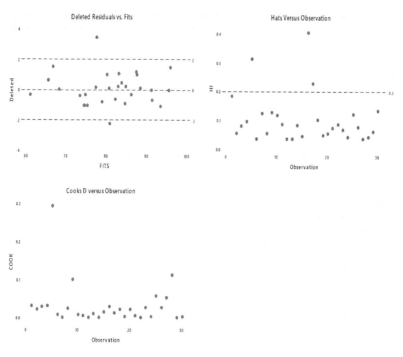

6.23 From Cooks D, we flag observations 5 and 28 as potential influential points. Observation 16 is also a potential leverage point based on the Hats vs. Observation plot and observation 28 is flagged as an outlier in the y-space on the Residuals vs. Predicted plot.

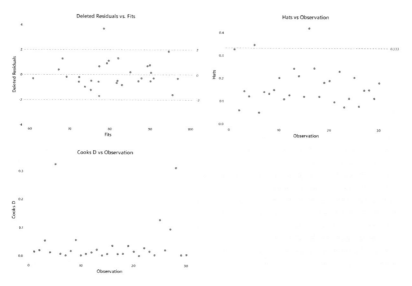

6.24 Observation 29 is being flagged as highly influential by the Cooks D statistic. Also, quite a few points are being flagged as leverage points in the Hats vs. Observation plot.

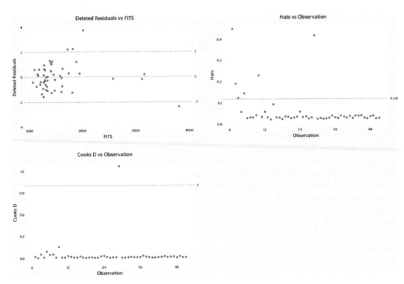

Chapter 7: Polynomial Regression Models

7.1   Yes there are potential problems since the correlation $(x, x^2) = .995$

7.2   a. $\widehat{y} = 1.63 - 1.23x + 1.49x^2$.

       b. $F = 1.86 \times 10^6$ with $p = 0.000$ which is significant.

       c. $F = \dfrac{4.607}{0.000} \approx \infty$ which is significant.

       d. Since it is a quadratic model, there can be potential hazards in extrapolating.

7.3   There is a problem with normality. The residuals seem to show that the model is adequate.

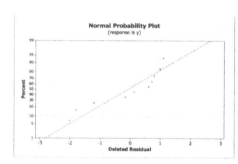
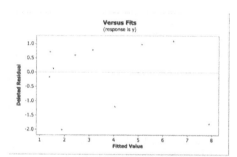

7.4   a. $\widehat{y} = -4.5 + 1.38x + 1.47x^2$.

       b. $F = 1044.99$ with $p = 0.000$ which is significant.

       c. $F = 48.7$ with $p = 0.001$ which is indicates lack of fit.

       d. $F = \dfrac{24.3}{2.7} = 9$ which is significant and indicates the term cannot be deleted.

7.5   There is an outlier which affects the normality and the residual plot which shows the model is not adequate.

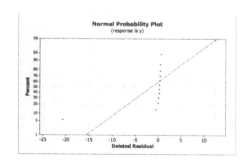
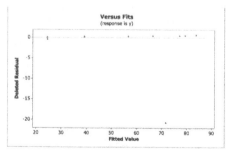

7.6 a. $\widehat{y} = 3025 - 194x_1 - 6.1x_2 + 3.63x_1^2 + 1.15x_2^2 - 1.33x_1x_2$.

b. $F = 177.17$ with $p = 0.000$ which is significant.

c. $F = .46$ with $p = .73$ which indicates there is no lack of fit.

d. $F = 2.21$ which is not significant and indicates that the interaction term does not contribute significantly to the model.

e. The quadratic term for x_2 contributes significantly to the model while the quadratic term for x_1 does not.

7.7 Observation 7 is influential which affects the plots. Normality looks pretty good and the residual plot is ok.

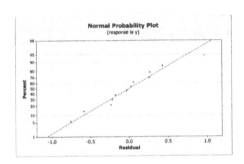
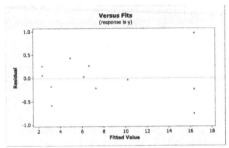

7.8 a. $\widehat{y} = 3.535 + .360P_1(x) + .187P_2(x)$.

b. $SS_R(\alpha_1, \alpha_2) = .360(118.71) + .187(24.66) = 47.31$. The linear and quadratic terms account for all of the variation in the data. Thus, the cubic term is not necessary.

7.9 a. To test $H_0 : \beta_{10} = \beta_{11} = \beta_{12} = 0$ use $F = \dfrac{SS_R(\beta_{10}, \beta_{11}, \beta_{12} | \beta_{00}, \beta_{01}, \beta_{02})/3}{MS_E}$.

b. Delete the term $\beta_{10}(x - t)^0$.

c. Also, delete the term $\beta_{11}(x - t)^1$.

7.10 A complete second-order model was fit to the delivery time data in Example 3.1. The analysis was done on centered data. Insignificant regressors were removed from the model.

The resulting regression equation is $\widehat{y} = 21.1 + 1.26 * (x_{num} - 8.76) + 0.0136 * (x_{dist} - 409.28) + 0.0306 * (x_{num} - 8.76)^2$.

Coefficient	test statistic	p-value
β_{num}	6.70	0.000
β_{dist}	4.36	0.000
β_{num}^2	2.98	0.007

The regression indicates that the quadratic term for the number of cases of product stocked improves the model.

7.11 A complete second order model was fit to the patient satisfaction data where the data have been centered.

The regression equation is $\hat{y} = 69.1 - 1.029 * (x_{age} - 50.84) - 0.422 * (x_{sev} - 45.92) + 0.0031 * (x_{age} - 50.84) * (x_{sev} - 45.92) - 0.0065 * (x_{age} - 50.84)^2 - 0.0082 * (x_{sev} - 45.92)$.

Coefficient	test statistic	p-value
β_{age}	-5.54	0.000
β_{sev}	-1.95	0.067
$\beta_{age*sev}$	0.14	0.892
β_{age}^2	-0.56	0.584
β_{sev}^2	-0.44	0.663

There is no indication that it is necessary to add these second-order terms to the model.

7.12 a. Change the ranges to $x \le t_1$, $t_1 < x \le t_2$, and $x > t_2$.

b. Delete the terms $\beta_{10}(x - t_1)^0$ and $\beta_{20}(x - t_2)^0$.

c. Also, delete the terms $\beta_{11}(x - t_1)^1$ and $\beta_{21}(x - t_2)^1$.

7.13 $\hat{y} = 15.1 - .0502x + .0389(x - 200)^1$. Test $H_0 : \beta_{11} = 0$, which gives a $t = 6.53$ and $p = 0.000$. The data do support the fit of this model.

7.14 $\hat{y} = 15.298 - .0516x + .325(x - 200)^0 + .0373(x - 200)^1$. Test $H_0 : \beta_{10} = 0$, which gives a $t = 0.79$ and $p = 0.456$. There is no change in the intercept but a change in the slope.

7.15 The variance inflation factors are 4.9 which do not indicate a multicollinearity problem.

7.16 a. The variance inflation factors are 19.9 which indicates there is a multicollinearity problem.

b. The variance inflation factors are 1.0 which indicates there is not a multicollinearity problem.

c. Many times centering can remove the multicollinearity problem.

7.17 a. The data are nonlinear.

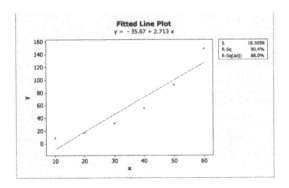

b. This also shows the data is nonlinear.

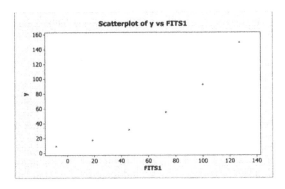

c. There is a quadratic pattern.

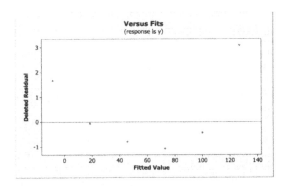

d. $\widehat{y} = 20.1 - 1.47x + .059x^2$. The test on the quadratic term is $F = \dfrac{1332.8}{12.5} = 106.62$ which is significant.

e. Yes, the second order model fits better.

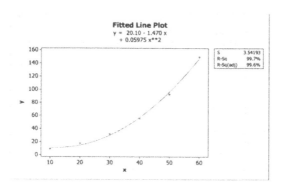

7.18 a. $\hat{y} = -1.77 + .421x_1 + .222x_2 - .128x_3 - .0193x_1^2 + .007x_2^2 + .0008x_3^2 - .019x_1x_2 + .009x_1x_3 + .003x_2x_3$.

   b. $F = 19.63$ with $p = 0.000$ which is significant. All are non-significant.

Coefficient	test statistic	p-value
β_1	1.43	0.172
β_2	1.70	0.108
β_3	−1.82	0.087
β_{11}	−1.15	.267
β_{22}	−.62	.545
β_{33}	.57	.575
β_{12}	−1.63	.118
β_{13}	1.20	.247
β_{23}	.37	.719

   c. There are several outliers which affect normality and the residual plot.

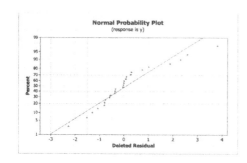
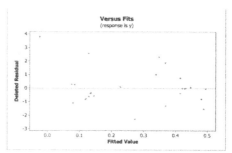

   d. $F = \dfrac{.035908/6}{.003712} = 1.61$ which is not significant.

7.19 The variance inflation factors are all very large indicating there is a serious problem with multicollinearity.

7.20 a. The predicted response at the point is $\widehat{y} = .2689$ and a 95% confidence interval on the mean response at the point is $(.2106, .3272)$.

b. The predicted response at the point is $\widehat{y} = .2512$ and a 95% confidence interval on the mean response at the point is $(.2185, .2840)$.

c. From the confidence intervals, it appears that the model without the pure quadratic terms might be better but the MS_{Res} are basically the same.

7.21 a. $\widehat{y} = -1709 + 2.02x - .00059x^2$.

b. $F = 300.11$ with $p = 0.000$ which is significant.

c. $F = \dfrac{2.428}{.044} = 55.18$ which is significant. Both terms should be included in the model.

d. There is a problem with normality and a possibility of nonconstant variance.

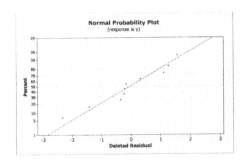
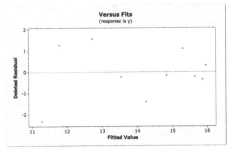

7.22 a. At $x = 1750$, the predicted response is $\widehat{y} = 14.8324$ and a 95% confidence interval on the mean response at the point is $(14.2841, 15.3808)$. At $x = 1775$, the predicted response is $\widehat{y} = 13.153$ and a 95% confidence interval on the mean response at the point is $(12.617, 13.6889)$.

b. At $x = 1750$, the predicted response is $\widehat{y} = 14.303$ and a 95% confidence interval on the mean response at the point is $(12.888, 15.718)$. At $x = 1775$, the predicted response is $\widehat{y} = 12.996$ and a 95% confidence interval on the mean response at the point is $(11.548, 14.444)$. The predicted values are closer to the actual values using the quadratic model. Also, the prediction intervals are shorter with the quadratic model.

7.23 The fitted model is $\widehat{y} = 44.976 + 4.339(x - 7.2632) - 0.5489(x - 7.2632)^2 - 0.05519(x - 7.2632)^3$. The observed value of $F_0 = 165.44$ and the p-value is small, so the hypothesis $H_0: \beta_1 = \beta_2 = \beta_3 = 0$ is rejected. We conclude that either the linear, quadratic, cubic, or some combination of the three contributed significantly to the model. The residual analysis does not reveal any serious model inadequacy. In terms of the test of $H_0: \beta_3 = 0$ vs. $H_1: \beta_3 \neq 0$

the value of $F_0 = 31.78$ with a small p-value. This cubic term contributes significantly to the model.

7.24 The report indicates an R^2 of 0.98, which is higher than the original quadratic polynomial fit R^2 of 0.91. These values are not directly comparable, but indicate we have a good fit to the data. The loess residual standard error is 2.177, compared to a value of 4.42 for the quadratic fit. This suggest that the loess is a better fit to the data. The loess requires an Equivalent Number of Parameters of 4.58, which are more parameters than the quadratic fit. Overall, the loess fit is preferred over the quadratic fit.

7.25 The cubic spline model is preferred over the loess model. The residual standard error for the loess model is 0.54 compared to 0.27 for the cubic spline model. The residual analysis for the loess fit also still suggests strong evidence of curvature that is no evident in the cubic spline fit.

7.26 a. $\widehat{y} = 63.78 + 2.8938\,(x - 17.875) - 0.07319(x - 17.875)^2$.

b. Yes, both the linear and quadratic terms are significant with sequential test statistics of $F_0 = 174.88$ for the linear term and $F_0 = 54.92$ for the quadratic term.

c. The residuals suggest that the quadratic polynomial is an adequate fit to the data.

d. For the cubic polynomial, the cubic term is insignificant with $F_0 = 0.01$. There is no substantial change in the R^2 and the MS_{Res} increased when the cubic term was added. This evidence suggests that the quadratic model is superior to the cubic model.

7.27 The loess fit results in a residual standard error of 2.797 and a $R^2 = 99$. The equivalent number of parameters is 4.57. The argument could be made that the loess fit is a better fit than the cubic polynomial fit. However, the quadratic fit is still the superior fit with a comparable residual standard error of 3.4, a similar R^2 value of 98.6. Also, the quadratic fit uses fewer parameters than both the loess fit and the cubic fit.

Chapter 8: Indicator Variables

8.1 β_0, β_2, β_3, and β_4 determine the intercept while the other parameters determine the slope.

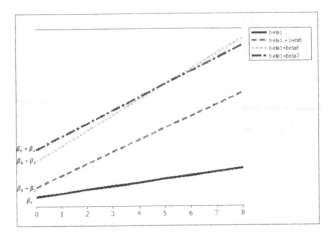

8.2 a.

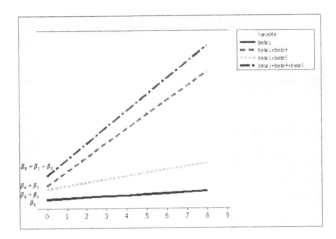

b.

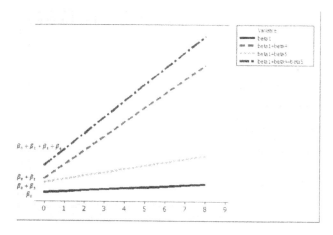

8.3   a. Let

$$x_3 = \begin{cases} 1 & \text{if San Diego} \\ 0 & \text{otherwise} \end{cases} \qquad x_4 = \begin{cases} 1 & \text{if Boston} \\ 0 & \text{otherwise} \end{cases} \qquad x_5 = \begin{cases} 1 & \text{if Austin} \\ 0 & \text{otherwise} \end{cases}$$

Then $\widehat{y} = .42 + 1.77x_1 + .011x_2 + 2.29x_3 + 3.74x_4 - .45x_5$.

b. No, $F = \dfrac{64.2/3}{8.9} = 2.41$ which is not significant.

c. There is a problem with normality and a pattern to the residuals.

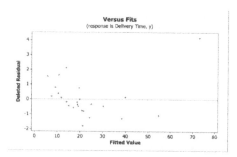

8.4   a. $\widehat{y} = 33.6 - .0457x_1 - .5x_{11}$. No, the $t = -.22$ with $p = 0.824$ which is not significant.

b. $\widehat{y} = 42.92 - .117x_1 - 13.46x_{11} + .082x_1x_{11}$. There is a significant interaction between engine displacement and the type of transmission. When the transmission is automatic, $\widehat{y} = (42.92 - 13.46) + (-.117 + .082)x_1 = 29.46 - .035x_1$ which indicates that on average for every increase of one cubic inch in displacement, miles per gallon decreases by .035. When the transmission is manual, $\hat{y} = 42.92 - .117x_1$ which indicates that on average for every increase of one cubic inch in displacement, miles per gallon decreases by .117.

8.5 a. $\hat{y} = 39.2 - .0048x_{10} - 2.7x_{11}$. No, the $t = -1.36$ with $p = 0.184$ which is not significant.

  b. $\hat{y} = 58.1 - .0125x_{10} - 26.2x_{11} + .009x_{10}x_{11}$. There is a significant interaction between vehicle weight and the type of transmission. When the transmission is automatic, $\hat{y} = (58.1 - 26.2) + (-.0125 + .009)x_{10} = 31.9 - .0035x_{10}$ which indicates that on average for every increase of one cubic inch in displacement, miles per gallon decreases by .0035. When the transmission is manual, $\hat{y} = 58.1 - .0125x_{10}$ which indicates that on average for every increase of one cubic inch in displacement, miles per gallon decreases by .0125.

8.6 Let
$$x_{51} = \begin{cases} 1 & \text{if } x_5 \text{ is negative} \\ 0 & \text{if } x_5 = 0 \end{cases} \qquad x_{52} = \begin{cases} 0 & \text{if } x_5 = 0 \\ 1 & \text{if } x_5 \text{ is positive} \end{cases}$$

This yields $\hat{y} = 19.4 - .007x_7 - .006x_8 + .46x_{51} + 2.33x_{52}$. The effect of turnovers is assessed by $F = \dfrac{22.276/2}{5.462} = 2.04$ which is not significant.

8.7 $E(y) = S(x) = \beta_{00} + \beta_{01}x_1 + \beta_{11}(x_1 - t)x_2$ where $x_2 = \begin{cases} 0 & \text{if } x_2 \leq t \\ 1 & \text{if } x_2 > t \end{cases}$.

8.8 $E(y) = S(x) = \beta_{00} + \beta_{01}x_1 + \beta_{10}x_2 + \beta_{11}(x_1 - t)x_2$ where $x_2 = \begin{cases} 0 & \text{if } x_2 \leq t \\ 1 & \text{if } x_2 > t \end{cases}$.

8.9
$$\mathbf{y} = \begin{pmatrix} y_{11} \\ y_{12} \\ y_{13} \\ y_{21} \\ y_{22} \\ y_{31} \\ y_{32} \\ y_{33} \\ y_{34} \\ y_{41} \\ y_{42} \\ y_{43} \end{pmatrix} \qquad \mathbf{X} = \begin{pmatrix} 1 & 1 & 0 & 0 \\ 1 & 1 & 0 & 0 \\ 1 & 1 & 0 & 0 \\ 1 & 0 & 1 & 0 \\ 1 & 0 & 1 & 0 \\ 1 & 0 & 0 & 1 \\ 1 & 0 & 0 & 1 \\ 1 & 0 & 0 & 1 \\ 1 & 0 & 0 & 1 \\ 1 & 0 & 0 & 0 \\ 1 & 0 & 0 & 0 \\ 1 & 0 & 0 & 0 \end{pmatrix}$$

No, $\hat{\beta}_0 = \overline{y}_{..} - \overline{y}_{1.} - \overline{y}_{2.} - \overline{y}_{3.} = \overline{y}_{4.}$, $\hat{\beta}_1 = \overline{y}_{1.} - \overline{y}_{4.}$, $\hat{\beta}_2 = \overline{y}_{2.} - \overline{y}_{4.}$, $\hat{\beta}_3 = \overline{y}_{3.} - \overline{y}_{4.}$.

8.10 a. $y_{1j} = \beta_0 + \beta_1 + \varepsilon_{1j}$, $y_{2j} = \beta_0 + \beta_2 + \varepsilon_{2j}$, $y_{3j} = \beta_0 - \beta_1 - \beta_2 + \varepsilon_{3j}$ which gives

$$\mu_1 = \beta_0 + \beta_1$$
$$\mu_2 = \beta_0 + \beta_2$$
$$\mu_3 = \beta_0 - \beta_1 - \beta_2.$$

Therefore, $\mu_1 + \mu_2 + \mu_3 = 3\beta_0$ implying that $\beta_0 = \dfrac{\mu_1 + \mu_2 + \mu_3}{3} = \bar{\mu}, \beta_1 = \mu_1 - \beta_0 = \mu_1 - \bar{\mu}$ and $\beta_2 = \mu_2 - \beta_0 = \mu_2 - \bar{\mu}..$

b.

$$\mathbf{y} = \begin{pmatrix} y_{11} \\ y_{12} \\ \vdots \\ y_{1n} \\ y_{21} \\ y_{22} \\ \vdots \\ y_{2n} \\ y_{31} \\ y_{32} \\ \vdots \\ y_{3n} \end{pmatrix} \qquad \mathbf{X} = \begin{pmatrix} 1 & 1 & 0 \\ 1 & 1 & 0 \\ \vdots & \vdots & \vdots \\ 1 & 1 & 0 \\ 1 & 0 & 1 \\ 1 & 0 & 1 \\ \vdots & \vdots & \vdots \\ 1 & 0 & 1 \\ 1 & -1 & -1 \\ 1 & -1 & -1 \\ \vdots & \vdots & \vdots \\ 1 & -1 & -1 \end{pmatrix}$$

c.

$$SS_R(\hat{\beta}_0, \hat{\beta}_1, \hat{\beta}_2) = \hat{\beta}'\mathbf{X}'\mathbf{y}$$

$$= \left(\bar{y}_{..} \quad \bar{y}_{1.} - \bar{y}_{..} \quad \bar{y}_{2.} - \bar{y}_{..} \right) \begin{pmatrix} y_{..} \\ y_{1.} - y_{3.} \\ y_{2.} - y_{3.} \end{pmatrix}$$

$$= y_{..}\bar{y}_{..} + (y_{1.} - y_{3.})(\bar{y}_{1.} - \bar{y}_{..}) + (y_{2.} - y_{3.})(\bar{y}_{2.} - \bar{y}_{..})$$

$$= (y_{1.} + y_{2.} + y_{3.})\bar{y}_{..} + y_{1.}(\bar{y}_{1.} - \bar{y}_{..}) + y_{2.}(\bar{y}_{2.} - \bar{y}_{..})$$

$$\quad - y_{3.}(\bar{y}_{1.} + \bar{y}_{2.} - 2\bar{y}_{..})$$

$$= y_{1.}\bar{y}_{1.} + y_{2.}\bar{y}_{2.} + y_{3.}(3\bar{y}_{..} - \bar{y}_{1.} - \bar{y}_{2.})$$

$$= y_{1.}\bar{y}_{1.} + y_{2.}\bar{y}_{2.} + y_{3.}\bar{y}_{3.}.$$

which is the same as the usual sum of squares.

8.11 a.

$$\mathbf{y} = \begin{pmatrix} 7 \\ 7 \\ 15 \\ 11 \\ 9 \\ 12 \\ 17 \\ 12 \\ 18 \\ 18 \\ 14 \\ 18 \\ 18 \\ 19 \\ 19 \\ 19 \\ 25 \\ 22 \\ 29 \\ 23 \\ 7 \\ 10 \\ 11 \\ 15 \\ 11 \end{pmatrix} \qquad \mathbf{X} = \begin{pmatrix} 1 & 1 & 0 & 0 & 0 \\ 1 & 1 & 0 & 0 & 0 \\ 1 & 1 & 0 & 0 & 0 \\ 1 & 1 & 0 & 0 & 0 \\ 1 & 1 & 0 & 0 & 0 \\ 1 & 0 & 1 & 0 & 0 \\ 1 & 0 & 1 & 0 & 0 \\ 1 & 0 & 1 & 0 & 0 \\ 1 & 0 & 1 & 0 & 0 \\ 1 & 0 & 1 & 0 & 0 \\ 1 & 0 & 0 & 1 & 0 \\ 1 & 0 & 0 & 1 & 0 \\ 1 & 0 & 0 & 1 & 0 \\ 1 & 0 & 0 & 1 & 0 \\ 1 & 0 & 0 & 1 & 0 \\ 1 & 0 & 0 & 0 & 1 \\ 1 & 0 & 0 & 0 & 1 \\ 1 & 0 & 0 & 0 & 1 \\ 1 & 0 & 0 & 0 & 1 \\ 1 & 0 & 0 & 0 & 1 \\ 1 & 0 & 0 & 0 & 0 \\ 1 & 0 & 0 & 0 & 0 \\ 1 & 0 & 0 & 0 & 0 \\ 1 & 0 & 0 & 0 & 0 \\ 1 & 0 & 0 & 0 & 0 \end{pmatrix}$$

b. $\widehat{\beta}_0 = 10.8$, $\widehat{\beta}_1 = -1$, $\widehat{\beta}_2 = 4.6$, $\widehat{\beta}_3 = 6.8$, $\widehat{\beta}_4 = 10.8$.

c. $\widehat{\beta}_1 - \widehat{\beta}_3 = -1 - 6.8 = -7.8$.

d. $F = 14.76$ with $p = 0.000$ which is significant and indicates that the mean tensile strength is not the same for all five cotton percentages.

8.12 a. Since $y_{ijk} = \mu + \tau_i + \gamma_j + (\tau\gamma)_{ij} + \varepsilon_{ijk}$ for $i = 1, 2$, $j = 1, 2$ and $k = 1, 2$, we get

$$y_{11k} = \mu + \tau_1 + \gamma_1 + (\tau\gamma)_{11} + \varepsilon_{11k}$$

$$y_{12k} = \mu + \tau_1 + \gamma_2 + (\tau\gamma)_{12} + \varepsilon_{12k}$$

$$y_{21k} = \mu + \tau_2 + \gamma_1 + (\tau\gamma)_{21} + \varepsilon_{21k}$$

$$y_{22k} = \mu + \tau_2 + \gamma_2 + (\tau\gamma)_{22} + \varepsilon_{22k}$$

Let

$$x_1 = \begin{cases} -1 & \text{if level 1 of treat type 1} \\ 1 & \text{if level 2 of treat type 1} \end{cases} \qquad x_2 = \begin{cases} -1 & \text{if level 1 of treat type 2} \\ 1 & \text{if level 2 of treat type 2} \end{cases}$$

Then, $y_{ijk} = \beta_0 + \beta_1 x_1 + \beta_2 x_2 + \beta_3 x_1 x_2 + \varepsilon_{ijk}$.

b.

$$\mathbf{y} = \begin{pmatrix} y_{111} \\ y_{112} \\ y_{121} \\ y_{122} \\ y_{211} \\ y_{212} \\ y_{221} \\ y_{222} \end{pmatrix} \qquad \mathbf{X} = \begin{pmatrix} 1 & -1 & -1 & 1 \\ 1 & -1 & -1 & 1 \\ 1 & -1 & 1 & -1 \\ 1 & -1 & 1 & -1 \\ 1 & 1 & -1 & -1 \\ 1 & 1 & -1 & -1 \\ 1 & 1 & 1 & 1 \\ 1 & 1 & 1 & 1 \end{pmatrix}$$

c. To test $H_0 : \tau_1 = \tau_2 = 0$ obtain the sum of squares for the first treatment type and form the ratio $F = \dfrac{MS_A}{MS_{Res}}$. Do the same for the other treatment type and the interaction.

8.13 a. $\widehat{y} = 8.32 + 1.12x_4 - 1.22r_1 - 2.76r_2$. The region does have an impact, $F = \dfrac{30.961/2}{.8} = 19.35$.

b. There is a slight departure from normality.

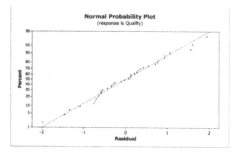
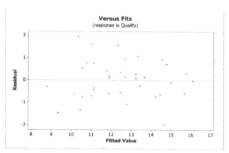

c. There are 2 outliers: observations 12 and 25.

d. $\widehat{y} = 10.1 + .796x_4 - 3.38r_1 - 6.28r_2 + .403x_4r_1 + .714x_4r_2$. No, the model is not superior to the model in part a.

8.14 The model in question 8.13 is superior.

Model	R^2	MS_{Res}	Region is Significant	Nonconstant Variance
Problem 8.13	80.9%	0.800	Yes	No
Problem 8.14	61.9%	1.584	No	Yes

8.15 Because LifeExp is the average between the male and female life expectancy, to predict average life, we can let

$$x_1 = \begin{cases} -1 & \text{if female} \\ 1 & \text{if male} \end{cases}$$

Also, recall from Problem 6.17 that a transformation was needed. If we again use the square roots of the response and the regressors, the model is $\widehat{y*} = 8.67 + 0.154x_1 - 0.0326sqrt(x_2) - 0.00704sqrt(x_3)$, with $F = 45.05$ and $p = 0.000$. $R^2 = 65.2\%$ and $R^2_{Adj} = 63.8\%$. $MS_{Res} = 0.0935$. Observations 8, 21, 30, 46, 59, and 68 are influential, as before, and considering this, there are no problems with the residual plots. This is very close to our model for average expected life from Problem 6.17: $\widehat{y*} = 8.67 - 0.0323sqrt(x_2) - 0.00713sqrt(x_3)$ with $MS_{Res} = 0.0902$ but includes the adjustment for gender, so all three responses can be fit with a single model.

8.16 The response variable INHIBIT was transformed by taking the square root due to problems with nonconstant variance in the original model. Let

$$x_2 = \begin{cases} 0 & \text{if Surface} \\ 1 & \text{if Deep} \end{cases}$$

The model is $\widehat{y*} = -0.264 + 121x_1 + 2.25x_2$. $F = 11.45$ with $p = 0.001$. $R^2 = 62.1\%$ and $R^2_{Adj} = 56.6\%$. No observations are influential, and the residual plots confirm the assumptions are not violated.

8.17 Adding the indicator variable has not improved the model. There is no evidence to support the claim that medical and surgical patients differ in their satisfaction as evident by the fact that the indicator variable is insignificant ($t = 0.48$ and $p - \text{value} = 0.633$).
      The regression equation is $\widehat{y} = 140 - 1.06x_{age} - 0.441x_{sev} + 1.99x_{sur-med}$.

Coefficient	test statistic	p-value
β_{age}	-6.51	0.000
β_{sev}	-2.42	0.025
$\beta_{sur*med}$	0.48	0.633

8.18 The addition of the indicator variable to the fuel consumption data does not seem to improve the analysis. In the analysis the only variable that significantly impacts fuel consumption is the initial boiling point x_5. The analysis below shows that adding the indicator variable is not a significant additional to the model. The proper analysis of these data is given in Exercise 5.20.

The regression equation is $\hat{y} = 413 - 4.25x_1 - 0.264x_5$.

Coefficient	test statistic	p-value
β_1	-1.09	0.295
β_5	-2.76	0.016

8.19 The model for the wine quality data was reduced to find significant predictors. The only significant predictor turned out to be wine color x_5. When the indicator for wine variety was added to the model, the variable was not significant at the 0.05 level with a $p - value = 0.17$. For this data we will also note that there was a strong problem with multicollinearity, so we are hesitant on the accuracy of this model.

The regression equation is $\hat{y} = 12 - 0.628x_1 + 0.850x_5$.

Coefficient	test statistic	p-value
β_1	-1.41	0.170
β_5	5.59	0.000

8.20 The regression for the methanol oxidation data was completed in Exercise 5.7. The indicator for reactor system was already included in the regression model. Exercise 5.7 concludes that the indicator variable is not significant for the transformed model.

8.21 a. The overall F statistic is 0.5138 with a p-value of 0.7646. There is no indication that any of the regressors have a major impact on PER. Also, the $R^2 = 0.046$ indicating that these regressors are not explaining any of the variability in PER.

   b. The overall F statistic is 0.79 with a p-value of 0.624. There is no indication that we have strengthened our regression by adding the player position to the model. The $R^2 = 0.127$, which is an increase. However, we would expect this because we have added more regressors to the model. The R^2 adjusted is 0%.

   c. The overall F statistic is 1.06 with a p-value of 0.434. There is no indication that we have strengthened our regression by adding the interactions between play position and the continuous variables to the model. The $R^2 = 0.515$, which is an increase. However, we would expect an increase because we have added more regressors to the model. The R^2 adjusted has gone up slightly to 0.03. This model is still not doing a good job at explaining PER.

8.22 Observations 1, 8, 40, and 57 were determined to be outliers or influential data. These observations were removed. The residual analysis of the assumptions is much more desirable. The overall F statistic is 1.77 with a p-value of 0.075. The $R^2 = 0.67$, which is an increase. Many of the model terms are not significant, so we still may be able to improve the model by removing some of the regressors.

8.23 After looking at both the models from 8.21 and 8.22, it was determined that standing vertical leap should be removed from the model. All interaction terms that included standing vertical leap were also removed. This resulted in a model with an overall F statistic of 2.39 and a p-value of 0.0125. The $R^2 = 0.66$ and the R^2 adjusted is 0.38. The value of R^2 adjusted has increased from 0.29 for the model fit in Problem 8.22. This increase is another justification from removing the predictor.

Chapter 9: Multicollinearity

9.1   a. The correlation between x_1 and x_2 is .824.

       b. The variance inflation factors are 3.1.

       c. The condition number of $\mathbf{X'X}$ is $\kappa = 40.68$ which indicates that multi-collinearity is not a problem in these data.

9.3   The eigenvector associated with the smallest eigenvalue is

Eigenvector
-0.839
0.081
0.437
0.117
0.289

All four factors contribute to multicollinearity.

9.5   There are two large condition indices in the non-centered data. In general, it is better to center.

Number	Condition Indices	x_1	x_2	x_3	x_4
1	1.000	.00037	.00002	.00021	.00004
2	7.453	.01004	.00001	.00266	.0001
3	14.288	.00058	.00032	.00159	.00168
4	109.412	.05745	.00278	.04569	.00088
5	62,290.176	.93157	.99687	.94985	.9973

9.7   a. The correlation matrix is

$$
\begin{array}{c c c c c c c c}
 & x_1 & x_2 & x_3 & x_6 & x_7 & x_8 & x_9 & x_{10} \\
x_2 & 0.945 & & & & & & & \\
x_3 & 0.989 & 0.964 & & & & & & \\
x_6 & 0.659 & 0.772 & 0.653 & & & & & \\
x_7 & -0.781 & -0.643 & -0.746 & -0.301 & & & & \\
x_8 & 0.855 & 0.797 & 0.864 & 0.425 & -0.663 & & & \\
x_9 & 0.801 & 0.718 & 0.788 & 0.316 & -0.668 & 0.885 & & \\
x_{10} & 0.946 & 0.883 & 0.943 & 0.521 & -0.718 & 0.948 & 0.902 & \\
x_{11} & 0.835 & 0.727 & 0.801 & 0.417 & -0.855 & 0.686 & 0.651 & 0.772 \\
\end{array}
$$

which indicates that there is a potential problem with multicollinearity.

b. The variance inflation factors are

Regressor	VIF
x_1	117.6
x_2	33.9
x_3	116.0
x_6	4.6
x_7	5.4
x_8	18.2
x_9	7.6
x_{10}	78.6
x_{11}	5.1

which indicates there is evidence of multicollinearity.

9.9   The condition indices are

1.00
9.65
61.93
126.11
2015.02
5453.08
44836.79
85564.32
5899200.59
8.86×10^{12}

which indicate a serious problem with multicollinearity.

9.11 The condition number is $\kappa = 24{,}031.36$ which indicates a problem with multicollinearity. The variance inflation factors shown below indicate evidence of multicollinearity.

Regressor	VIF
x_1	3.67
x_2	7.73
x_3	19.20
x_4	7.46
x_5	4.69
x_6	7.73
x_7	1.12

9.13 The condition number is $\kappa = 12400885.78$ which indicates a problem with multicollinearity. The variance inflation factors shown below indicate evidence of multicollinearity.

Coefficient	test statistic	p-value	VIF
β_1	-0.88	0.408	1.00
β_2	0.16	0.874	1.901
β_3	0.35	0.734	168.467
β_4	-0.19	0.854	43.104
β_5	-0.47	0.655	60.791
β_6	-0.12	0.911	275.473
β_7	-0.10	0.925	185.707
β_8	-0.23	0.822	44.363

9.15 The condition number is $\kappa = 286096.79$ which indicates a problem with multicollinearity. The variance inflation factors shown below indicate evidence of multicollinearity, especially x_2 and x_3.

Coefficient	test statistic	p-value	VIF
β_1	3.09	0.009	1.519
β_2	5.70	0.000	26.284
β_3	3.91	0.002	26.447
β_4	0.21	0.840	2.202
β_5	-0.21	0.833	1.923

9.17 a. Using $k = .008$ gives a model with $R^2 = 97.8\%$ and $\sqrt{MS_{Res}} = .041$.

   b. Without the use of ridge regression it is .0196 and with ridge regression it is .0218, which is an increase of about 11%.

   c. Both are good models.

9.19 a. The ridge trace leads to $k = .18$, but the resulting model is not adequate.

   b. Without the use of ridge regression it is 0.00104 and with ridge regression it is 0.56265, which is an increase of about 540%.

   c. Without the use of ridge regression it is 99.2% and with ridge regression it is 43.7%, which is an decrease of over 50%.

9.21 a. Principal components regression yields $R^2 = 96.5\%$ while least squares yields $R^2 = 98.2\%$. The loss is minimal at around 2%.

   b. The coefficient vector is reduced to one term.

   c. The principal components model has virtually the same R^2 but has a higher $SS_E = 0.0351$ compared to the $SS_E = 0.0218$ with the ridge model.

9.23 a. The variance inflation factors are given below.

Regressor	VIF
PRECIP	2.0
EDUC	1.5
NONWHITE	1.3
NOX	1.7
SO2	1.4

The correlation matrix is

$$
\begin{array}{l}
\quad\quad\quad\quad\quad\quad PREC \quad EDUC \quad NONWHITE \quad NOX \\
\begin{array}{l}
EDUC \\
NONWHITE \\
NOX \\
SO2
\end{array}
\left(
\begin{array}{rrrr}
-0.490 & & & \\
0.403 & -0.209 & & \\
-0.486 & 0.230 & 0.025 & \\
-0.107 & -0.234 & 0.162 & 0.412
\end{array}
\right)
\end{array}
$$

There is no evidence of multicollinearity.

b. The ridge trace shows flat lines.

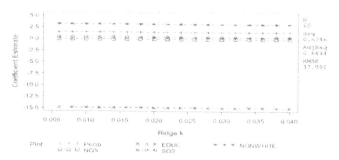

c. The ridge trace indicates $k = 0$, therefore the estimates of the coefficients for ridge and OLS are the same.

d. Principle-component regression gives

Eigenvalue	1.9648	1.4736	0.8348	0.4062	0.3206
Proportion	0.393	0.295	0.167	0.081	0.064
Cumulative	0.393	0.688	0.855	0.936	1.000

Variable	PC1	PC2	PC3	PC4	PC5
PRECIP	-0.641	0.007	0.093	0.038	0.761
EDUC	0.490	-0.305	0.551	0.510	0.323
NONWHITE	-0.345	0.410	0.750	0.011	-0.387
NOX	0.471	0.484	0.167	-0.596	0.401
SO2	0.095	0.710	-0.312	0.619	0.080

The principal components regression accounts for 85.5% of the variation with three variables while OLS (and ridge regression since k=0) accounts for only 67.5% of the variation in the model with five variables.

9.25 The shrinkage is on the scale versus the location.

9.27 You cannot find the k that minimizes $E(L_1^2)$ because the k does not depend on j. Thus the sums will not collapse making it impossible to isolate k (see problem 9.24).

9.29 Attempting to shrink only the independent variables that are contributing to the multicollinearity instead of shrinking the entire vector of independent variables will introduce less bias. However, shrinking only a subset of the regressors can create new problems and one must be sure of the subset they are choosing to shrink. It is still better to use ordinary ridge regression.

Chapter 10: Variable Selection and Model Building

10.1 a. With $\alpha = 0.10$, the model chosen is $y = \beta_0 + \beta_2 x_2 + \beta_7 x_7 + \beta_8 x_8 + \varepsilon$.

   b. With $\alpha = 0.10$, the model chosen is $y = \beta_0 + \beta_2 x_2 + \beta_7 x_7 + \beta_8 x_8 + \varepsilon$.

   c. With $\alpha_{IN} = 0.05$ and $\alpha_{OUT} = 0.10$, the model chosen is $y = \beta_0 + \beta_2 x_2 + \beta_7 x_7 + \beta_8 x_8 + \varepsilon$.

   d. The three procedures chose the same model.

10.3 The choice of cut-off values is to prevent the circular addition-subtraction of the variables.

10.5 a. The model involves just x_1 with $R_p^2 = 75.3\%$, $C_p = -1.8$ and $\sqrt{MS_{Res}} = 3.12$.

   b. Stepwise leads to the same model involving just x_1.

10.7 When $F_{IN} = F_{OUT} = 4.0$, the model involves only x_6. However, when $F_{IN} = F_{OUT} = 2.0$, the model involves x_6 and x_7.

10.9 The model involves x_1, x_2, x_3, and x_4 with $R_p^2 = 95.3\%$, $C_p = 5$ and $\sqrt{MS_{Res}} = .002$.

10.11 The model is $y = \beta_0 + \beta_1 x_1 + \beta_2 x_2 + \beta_4 x_4$ which has a PRESS $= 85.35$ and $R_{Pred}^2 = 96.86\%$.

10.13 a. From Section 3.7, we get $\widehat{\beta} = (\mathbf{W'W})^{-1}\mathbf{W'y}^0$ with $\mathbf{W'W} = \begin{pmatrix} 1 & r_{12} \\ r_{12} & 1 \end{pmatrix}$.

   Therefore, $(\mathbf{W'W})^{-1} = \begin{pmatrix} \dfrac{1}{1-r_{12}^2} & \dfrac{-r_{12}}{1-r_{12}^2} \\ \dfrac{-r_{12}}{1-r_{12}^2} & \dfrac{1}{1-r_{12}^2} \end{pmatrix}$ which means that

   $Var(\widehat{\beta_1^*}) = \dfrac{\sigma^2}{1-r_{12}^2}$.

   b. Since we are fitting a model with only one regressor, the $\mathbf{W'W}$ is the scalar 1. Thus its inverse is also the scalar 1 and the $Var(\widehat{\beta_1}) = \sigma^2$.

   c. We have seen from problem 3.31 earlier that in general $E(\widehat{\beta_1}) = \beta_1 + (\mathbf{X_1'X_1})^{-1}\mathbf{X_1'X_2}\beta_2$. For this problem, we have only 2 parameters and we are using the correlation form of the variables. Thus, $E(\widehat{\beta_1}) = \beta_1 + r_{12}\beta_2$ since the $\mathbf{W}_1$'s are the scalar 1 and $\mathbf{W}_2$ is the scalar r_{12}.

   d. $MSE = Var(\widehat{\beta_1}) + [E(\widehat{\beta_1}) - \beta_1]^2 = \sigma^2 + r_{12}^2\beta_2^2$. For $\widehat{\beta_1}$ to be preferable, we need $MSE(\widehat{\beta_1}) < Var(\widehat{\beta_1^*})$ which can be written as satisfying $\beta_2^2 < \dfrac{\sigma^2}{1-r_{12}^2}$.

10.15 Stepwise produces the model with x_4, x_5, r_1 and r_2 which is the model with the lowest C_p from 9.14 part a.

10.17 The model with SOAKTIME and DIFFTIME is selected with $C_p = 2.8$. There is a slight departure from normality and the several outliers and influential points.

10.19 The model with DIFFTIME, x_1 and x_2 is selected with $C_p = 4.7$. There is still a slight departure from normality and the several outliers and influential points.

10.21 The model with x_1, x_3, x_5 and x_6 is selected with $C_p = 5.6$. There is a departure from normality in the tails but the residual plot show the model is adequate. There are a couple of outliers.

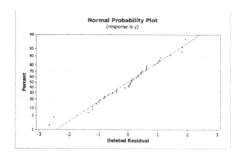 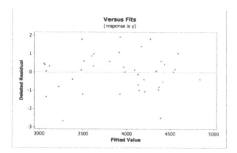

10.23 The confidence intervals for the model in 10.21 are narrower than those for 10.22. Also, the value for the PRESS statistic is smaller, 31685.4 compared to 34081.6.

10.25 Stepwise produces the same model as 10.24.

10.27 a. The model with x_1, x_2, x_3 and x_4 is selected ($C_p = 4.3$) which is the same model as in 10.24.

   b. Stepwise produces the same model.

   c. The confidence intervals for the new data set without observation 2 are narrower than the one from 10.26. The large residual from observation 2 increased MS_{Res} which in turn widened the confidence intervals in 10.26.

10.29 a. As in Section 10.4, we will use the log of the response and the log of viscosity for the model. For Run 0, performing best subsets produces the following table.

Variables	R-sq	R-sq (adj)	Mallow's Cp	S	Log (x_1)	x_2	x_3	x_5	x_6
1	64.7	62.2	6.7	0.12703	X				
1	20.1	14.4	30.2	0.19113		X			
2	75.2	71.4	3.1	0.11043	X	X			
2	67.6	62.6	7.1	0.12635	X		X		
3	78.3	72.8	3.5	0.10769	X	X			X
3	76.5	70.7	4.4	0.11191	X	X		X	
4	80.2	73.0	4.5	0.10737	X	X		X	X
4	79.3	71.8	4.9	0.10972	X	X	X		X
5	81.1	71.6	6.0	0.11006	X	X	X	X	X

From this, we would choose the model with 3 variables, $Log(x_1)$, x_2, and x_6, which are Log(Visc), Surface, and Voids, respectively. This gives the prediction equation $\widehat{Log(y)} = -1.54 - 0.507Log(x_1) + 0.454x_2 + 0.109x_6$.

For Run = 1, we get the following table from best subsets regression.

Variables	R-sq	R-sq (adj)	Mallow's Cp	S	Log (x_1)	x_2	x_3	x_5	x_6
1	59.7	56.6	6.3	0.16536	X				
1	21.6	15.5	22.6	0.23060			X		
2	71.3	66.5	3.3	0.14527	X				X
2	68.4	63.1	4.6	0.15239	X		X		
3	77.2	71.0	2.8	0.13522	X		X		X
3	74.6	67.6	3.9	0.14278	X			X	X
4	78.1	69.3	4.4	0.13901	X	X	X		X
4	77.8	68.9	4.5	0.13999	X		X	X	X
5	79.0	67.4	6.0	0.14335	X	X	X	X	X

From this, we choose a 3-variable model including $Log(x_1)$, x_3, and x_6, which are Log(Visc), base, and voids, respectively. The prediction equation is $\widehat{Log(y)} = -2.06 - 0.613Log(x_1) + 0.485x_3 + 0.187x_6$.

b. When we look at the separate runs, we see different regressors are most appropriate. While Log(Visc) and Voids are significant in both models, the percentage of asphalt in the surface course (x_2) is significant only in the first run ($x_4 = 0$). Also, the percentage of asphalt in the base course (x_3) is significant in the second run ($x_4 = 1$) but not the first.

c.

Model	R^2_{Adj}	MS_{Res}	Cp
Run $= 0$	72.8%	0.01160	3.5
Run $= 1$	71.0%	0.01828	2.8
Section 9.4	95.3%	0.09150	2.9

The model in Section 10.4 has more predictive power, but greater error than the models created for the two runs. Because the indicator variable for Run (x_4) was determined to not be significant in the model in Section 10.4, we would not expect an advantage in modeling the runs separately, other than this decrease in error.

10.31 The model with age and severity is selected ($C_p = 2.0$) from the all-possible-regressions selection. This same modal is selected from stepwise regression. An analysis of this data can be found in Section 3.6.

10.33 The all-possible-regressions selection on the wine quality of young red wines produced multiple candidate models. We first chose to look at a 6 regressor model (x_1, x_2, x_3, x_4, x_5, x_8) with a $C_p = 5.7$, $R^2 = 66.6$, $R^2_{adj} = 58.5$, and $s = 1.1403$.

The VIF's still indicate a problem with multicollinearity between x_4 and x_5. Without any advice from a subject matter expert, the decision was made to remove x_4 from the model. This results in a slight increase in s, but this is preferred since the model no longer suffers from the multicollinearity problem. The residual analysis does not indicate any problems with model adequacy.

Stepwise regression suggested the simple linear regression model only containing x_5. The fit criteria for this model include $s = 1.27181$, $R^2 = 50.1\%$ and $R^2_{adj} = 48.4\%$.

10.35 For the Hald cement data, LASSO shrinks the model parameter for x_3 (tetracalcium alumino ferrite) to zero. The suggested model from LASSO is the one to include x_1, x_2, and x_4. This was one of the candidate models explored based on the all possible regressions procedure. This model has the lowest value for MS_{Res}. The final model suggested in Chapter 10 differed slightly from the LASSO model. The model dropped x_4 and only included x_1 and x_2 because of the multicollinearity between x_2 and x_4.

10.37 The model that includes the predictors runs, ERA, and Errors is selected by the stepwise selection procedure. The residual analysis (similar to the one completed in Chapter 4) indicates no real problems with model adequacy other than the fact that the Texas Rangers are a potential outlier in the dataset. This model is one of the candidate models suggested by the all possible regressions. This is the best 3 predictor model and is competitive with the best 2 predictor model which includes runs and runs allowed per game (RA/G).

10.39   The model that includes max vertical leap is the suggested model from the stepwise selection method. This is also the suggested model from the all possible regressions model. However, it is important to note that there is room to improve this model with a R^2 value of only 4.09%. The residual analysis does not show any problems with model adequacy. However, it is suggested that observation 1 is a potential outlier.

10.41   The model suggested by the stepwise regression procedure includes population, 95th percentile income, and median price/sqft as the predictors. This model has a high R^2 at 91.58%. This is also the model suggested by the all possible regression procedure. The residual analysis indicates some concern with constant variance and potential outliers.

10.43   The model suggested by the stepwise regression procedure includes four regressors: off the tee, approach to the green, around the green, and putting. This is also the model suggested by the all possible regression procedure. The residual analysis indicates no problems with model adequacy.

Chapter 11: Validation of Regression Models

11.1 a. PRESS = 87.4612 with

$$R^2_{Pred} = 1 - \frac{\text{PRESS}}{SS_T}$$

$$= 1 - \frac{87.4612}{326.964}$$

$$= 73.25\%$$

The predictive power is not bad.

b. $\widehat{y} = -8.5.004x_2 + .28x_7 - .005x_8$

y	$\widehat{y}$
10	5.83
11	8.84
11	12.07
4	0.73
10	7.46
5	2.82

The model does not predict very well.

c. The model does a good job predicting these observations.

City	y	$\widehat{y}$
Dallas	11	10.71
Los Angeles	10	12.25
Houston	5	5.29
San Francisco	8	8.42

11.3 PRESS = 70.82 with $R^2_{Pred} = 59.5\%$ which agrees with problem 11.2 that the model does not predict well.

11.5 a. $\widehat{y}_p = 4.42 + 1.53x_1 + .012x_2$.

b. $\widehat{y}_e = 3.51 + 1.39x_1 + .016x_2$. The models are similar which indicates the overall model should be valid.

c. The model predicts fairly well and is consistent with Example 11.3.

11.7   PRESS = 337.37 with $R^2_{Pred} = 72.74\%$ which indicates that the predictive performance of the model is not bad.

11.9   The model is not predicting very well.

y	$\widehat{y}$
18.9	12.43
18.25	14.52
34.7	23.09
36.5	22.33
14.89	5.26
16.41	16.89
13.9	11.67
20.0	19.22

11.11 The standard errors are larger in the estimation set.

Problem 15.11		Problem 3.5	
Coefficient	Standard Error	Coefficient	Standard Error
$\widehat{\beta}_0$	2.409	$\widehat{\beta}_0$	1.535
$\widehat{\beta}_1$	0.009	$\widehat{\beta}_1$	0.006
$\widehat{\beta}_2$	0.936	$\widehat{\beta}_2$	0.671

11.13 The DUPLEX algorithm is probably not efficient for large sample sizes since $\binom{n}{2}$ is going to very large.

11.15 From Appendix C, we get

$$(\mathbf{X}'_{(i)}\mathbf{X}_{(i)})^{-1} = (\mathbf{X}'\mathbf{X})^{-1} + \frac{(\mathbf{X}'\mathbf{X})^{-1}\mathbf{x}_i\mathbf{x}'_i(\mathbf{X}'\mathbf{X})^{-1}}{1 - h_{ii}}$$

If we postmultiply the above by $\mathbf{x}_i$ we get

$$(\mathbf{X}'_{(i)}\mathbf{X}_{(i)})^{-1}\mathbf{x}_i = (\mathbf{X}'\mathbf{X})^{-1}\mathbf{x}_i + \frac{(\mathbf{X}'\mathbf{X})^{-1}\mathbf{x}_i\mathbf{x}'_i(\mathbf{X}'\mathbf{X})^{-1}\mathbf{x}_i}{1 - h_{ii}}$$

$$= (\mathbf{X}'\mathbf{X})^{-1}\mathbf{x}_i + \frac{(\mathbf{X}'\mathbf{X})^{-1}\mathbf{x}_i h_{ii}}{1 - h_{ii}}$$

$$= \frac{(\mathbf{X}'\mathbf{X})^{-1}\mathbf{x}_i[1 - h_{ii} + h_{ii}]}{1 - h_{ii}}$$

$$= \frac{(\mathbf{X}'\mathbf{X})^{-1}\mathbf{x}_i}{1 - h_{ii}}$$

Now, we will postmultiply the result from Appendix C by $\mathbf{X}'\mathbf{y}$

$$(\mathbf{X}'_{(i)}\mathbf{X}_{(i)})^{-1}\mathbf{X}'y = (\mathbf{X}'\mathbf{X})^{-1}\mathbf{X}'y$$

$$+ \frac{(\mathbf{X}'\mathbf{X})^{-1}\mathbf{x}_i\mathbf{x}'_i(\mathbf{X}'\mathbf{X})^{-1}\mathbf{X}'y}{1 - h_{ii}}$$

$$(\mathbf{X}'_{(i)}\mathbf{X}_{(i)})^{-1}[\mathbf{X}'_{(i)}y_{(i)} \ \mathbf{x}_iy_i] = \hat{\beta} + \frac{(\mathbf{X}'\mathbf{X})^{-1}\mathbf{x}_i\mathbf{x}'_i\hat{\beta}}{1 - h_{ii}}$$

$$(\mathbf{X}'_{(i)}\mathbf{X}_{(i)})^{-1}\mathbf{X}'_{(i)}y_{(i)} + (\mathbf{X}'_{(i)}\mathbf{X}_{(i)})^{-1}\mathbf{x}_iy_i = \hat{\beta} + \frac{(\mathbf{X}'\mathbf{X})^{-1}\mathbf{x}_i\hat{y}_i}{1 - h_{ii}}$$

$$\hat{\beta}_{(i)} + \frac{(\mathbf{X}'\mathbf{X})^{-1}\mathbf{x}_iy_i}{1 - h_{ii}} = \hat{\beta} + \frac{(\mathbf{X}'\mathbf{X})^{-1}\mathbf{x}_i\hat{y}_i}{1 - h_{ii}}$$

$$\hat{\beta}_{(i)} = \hat{\beta} + \frac{(\mathbf{X}'\mathbf{X})^{-1}\mathbf{x}_i\hat{y}_i}{1 - h_{ii}} - \frac{(\mathbf{X}'\mathbf{X})^{-1}\mathbf{x}_iy_i}{1 - h_{ii}}$$

$$\hat{\beta}_{(i)} = \hat{\beta} + \frac{(\mathbf{X}'\mathbf{X})^{-1}\mathbf{x}_i(\hat{y}_i - y_i)}{1 - h_{ii}}$$

$$\hat{\beta}_{(i)} = \hat{\beta} - \frac{(\mathbf{X}'\mathbf{X})^{-1}\mathbf{x}_i\hat{e}_i}{1 - h_{ii}}$$

11.17 a. The model is $y = \beta_0 + \beta_1 x_1 + \beta_4 x_4 + \beta_5 x_5 + \beta_6 x_6$ with $C_p = 5.0$.

b. The fitted model is $\hat{y} = -302 + 1.11x_1 + 5.32x_4 + 1.56x_5 - 13.3x_6$ with $R^2_{Pred} = 99.62\%$ which indicates the model is adequate and predicts very well.

11.19 a. $R^2_{Pred} = 75.81\%$

b. $\overline{R}^2_{Pred} = 77.45\%$

c. $R^2_{Pred} = 78.04\%$

d. All three parts produce relatively the same value for R^2_{Pred}.

Chapter 12: Introduction to Nonlinear Regression

12.1   As θ_2 decreases, the curve becomes steeper.

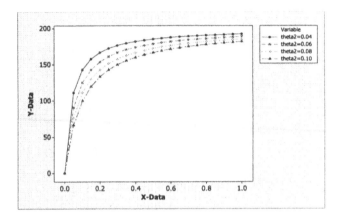

12.3   As θ_3 increases, the curve becomes steeper.

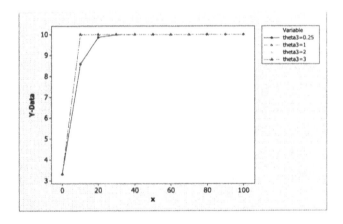

12.5   a. As θ_2 decreases, the curve becomes steeper.

   b. As $x \to \infty$, $E(y) \to 1$.

c. When $x = 0$, $E(y) = \theta_1 \exp\{-\theta_2\}$.

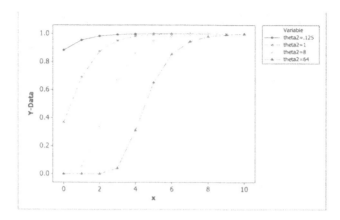

12.7 a. This is an intrinsically linear model.

$$y = [\theta_1 e^{\theta_2 + \theta_3 x}]\varepsilon$$
$$\ln(y) = \ln(\theta_1) + \theta_2 + \theta_3 x + \ln(\varepsilon)$$
$$y* = (\theta_1^* + \theta_2) + \theta_3 x + \varepsilon*$$

b. The model is nonlinear.

c. The model is nonlinear.

d. This is an intrinsically linear model.

$$y = [\theta_1 (x_1)^{\theta_2} (x_2)^{\theta_3}]\varepsilon$$
$$\ln(y) = \ln(\theta_1) + x_1 \ln(\theta_2) + x_2 \ln(\theta_3) + \ln(\varepsilon)$$
$$y* = \theta_1^* + x_1 \theta_2^* + x_2 \theta_3^* + \varepsilon*$$

e. The model is nonlinear.

12.9 $\hat{y} = -.121 x_2 + 1.066 e^{.4928 x_1}$. An approximate 95% confidence interval for θ_3 is $(-1.027, .785)$. Since this interval contains 0, we conclude there is no difference in the two days.

12.11 a.

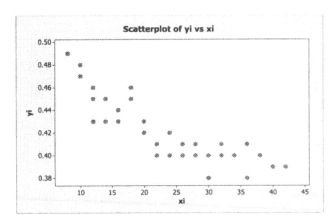

b. $\widehat{y} = .3896 - (-.2194)e^{-.0992x}$. The starting values were obtained by plotting the expectation function.

c. $F = 141.55$ with $p = < 0.0001$ which is significant.

d. An approximate 95% confidence interval for θ_1 is $(.3778, .4014)$. An approximate 95% confidence interval for θ_2 is $(-.2828, -.1560)$. An approximate 95% confidence interval for θ_3 is $(.0626, .1357)$. θ_2 is not different from zero.

e. The residuals show that the model is adequate.

12.13 a. $\widehat{y} = .8703(x_1)^{.783}(x_2)^{.227}$.

b. $F = 422.93$ with $p = 0.000$ which is significant. Both variables appear to have important effects.

c. The residual plots are better than in 12.12. The model seems adequate.

d. The nonlinear model.

12.15 a. There is a nonlinear pattern.

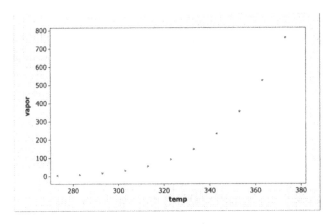

b. There is a problem with normality and a nonlinear pattern in the residuals. The regression equation is $vapor = -1956 + 6.69temp$.

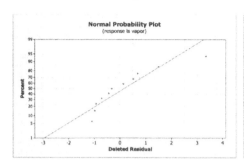

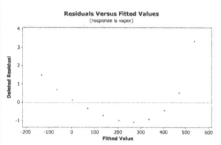

c. There is a slight improvement in the model. However, there contiues to be a problem with normality and a nonlinear pattern in the residuals. The new regression equation is $ln(vapor) = 20.6074 - 5200.76(1/temp)$

d. The appropriate nonlinear model is $vapor = \theta_0 e^{\theta_1(1/temp)}$. The estimated coefficients are $\widehat{\theta_0} = 576741131$ and $\widehat{\theta_1} = -5050$. We still notice a pattern in the residuals.

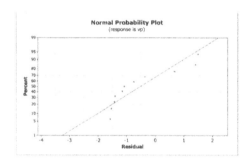

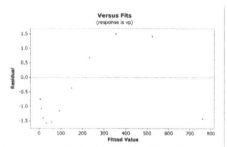

Note: To determine starting values, the nonlinear equation was linearized and the estimates from simple linear regression on a subset of the data were used as starting values. Another way for determining the starting value for θ_1 would be to use the chemical theory that the heat of vaporization (H_v) for water is $H_v = 9729\,cal/mole$. The ideal gas constant (R) is $R = 1.9872\,cal/mole^{\circ}K$. Therefore, a starting value for θ_1 is $\dfrac{H_v}{R} = 4895.8$.

e. The simple linear regression models differ from the nonlinear model in terms of the error structures. We prefer the nonlinear model because it appears to be a better fit to the data. However, there is a still a problem with the residuals because the chemical theory assumes an idea gas and that assumption is violated with real data.

12.17 a. The fitted model is $\hat{y} = 22.5 + 0.0228w - 0.01666d$. The residual analysis outlines one outlying point that should be investigated.

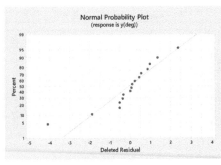
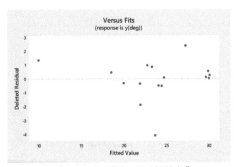

b. The fitted nonlinear model is $\hat{y} = 32.46\left[1 - \exp\left(-1.51\left(\dfrac{w}{d}\right)\right)\right]$. Note the starting values used were 16 for β_1 and 0.3314 for β_2. The residual analysis shows that the model is adequately fit to the data.

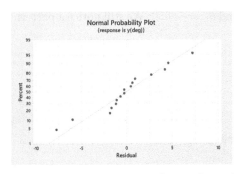
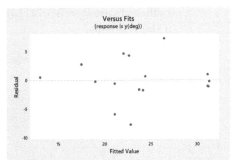

c. We prefer the nonlinear fit to the model because it appears to be a better fit to the data in terms of the residual analysis. The outlying point in the residual analysis for the linear fit to the data is not seen in the residual analysis of the nonlinear model fit and there is more random scatter in the Residual versus Fits graph.

12.19 a. The estimates are $\hat{\beta}_1 = 34.00$ and $\hat{\beta}_2 = 0.6271$. The starting estimates used were $\hat{\beta}_1 = 22$ and $\hat{\beta}_2 = 0.729$.

b. The standard error for the parameter estimates are $SE[\hat{\beta}_1] = 0.955$ and $SE[\hat{\beta}_2] = 0.077$.

Chapter 13: Generalized Linear Models

13.1 a. $\widehat{\pi} = \dfrac{1}{1 + e^{(-6.07 + 0.177x)}}$

   b. Deviance $= 17.59$ with $p = 0.483$ indicating that the model is adequate.

   c. $\widehat{O}_R = e^{-.0177} = .9825$ indicating that for every additional knot in speed the odds of hitting the target decrease by 1.75%.

   d. The difference in the deviances is basically zero indicating that there is no need for the quadratic term.

13.3 a. $\widehat{\pi} = \dfrac{1}{1 + e^{(-5.34 + .0015x)}}$

   b. Deviance $= .372$ with $p = 1.000$ indicating that the model is adequate.

   c. The difference in the deviances is $\mathrm{Dev}(x) - \mathrm{Dev}(x, x^2) = .372 - .284 = .088$ indicating that there is no need for the quadratic term.

   d. For $H_0 : \beta_1 = 0$, the Wald statistic is $Z = -.42$ which is not significant. For $H_0 : \beta_2 = 0$, the Wald statistic is $Z = -.30$ which is not significant.

   e. An approximate 95% confidence interval for β_1 is $(-.0018, .0033)$ and an approximate 95% confidence interval for β_2 is $(7.15 \times 10^{-7}, 5.27 \times 10^{-7})$.

13.5 a. $\widehat{\pi} = \dfrac{1}{1 + e^{(12.35 - .0002x_1 - 1.259x_2)}}$

   b. Deviance $= 14.76$ indicating that the model is adequate.

   c. For $\widehat{\beta}_1$, we get $\widehat{O}_R \approx 1$ indicating that the odds are basically even. For $\widehat{\beta}_1$, we get $\widehat{O}_R = 3.52$ indicating that every one year increase in the age of the current car increases the odds of purchasing a new car by 252%.

   d. $\widehat{\pi} = \dfrac{1}{1 + e^{(12.35 - .0002(45000) - 1.259(5))}} = .76$

   e. The difference in the deviances is $\mathrm{Dev}(x_1, x_2) - \mathrm{Dev}(x_1, x_2, x_1 x_2) = 14.764 - 10.926 = 3.838$ indicating that the interaction term could be included.

   f. For $H_0 : \beta_1 = 0$, the Wald statistic is $Z = -.26$ which is not significant. For $H_0 : \beta_2 = 0$, the Wald statistic is $Z = -.80$ which is not significant. For $H_0 : \beta_{12} = 0$, the Wald statistic is $Z = 1.13$ which is not significant.

   g. An approximate 95% confidence interval for β_1 is $(-.0005, .0004)$, an approximate 95% confidence interval for β_2 is $(-10.827, 4.555)$ and an approximate 95% confidence interval for β_{12} is $(-.0001, .0003)$.

13.7 a. $\widehat{\pi} = e^{(-3.61 - .0014x_1 + .0626x_2 - .0021x_3 - .0289x_4)}$

   b. Deviance $= 37.92$ indicating that the model is adequate.

   c. This indicates that x_3 should be removed.

d. Consider $\alpha = 0.05$ for all tests. For $H_0 : \beta_1 = 0$, the Wald statistic is $Z = 1.73$ which is not significant. For $H_0 : \beta_2 = 0$, the Wald statistic is $Z = 5.08$ which is significant. For $H_0 : \beta_3 = 0$, the Wald statistic is $Z = .13$ which is not significant. For $H_0 : \beta_4 = 0$, the Wald statistic is $Z = 1.87$ which is not significant.

e. An approximate 95% confidence interval for β_1 is $(-.0031, .0002)$, an approximate 95% confidence interval for β_2 is $(.0384, .0867)$, an approximate 95% confidence interval for β_3 is $(-.012, .0079)$, and an approximate 95% confidence interval for β_4 is $(-.0592, .0014)$.

13.9   Normality seems to be satisfied but there is a pattern to the residuals.

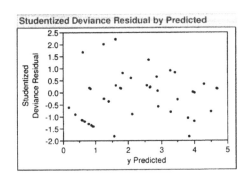
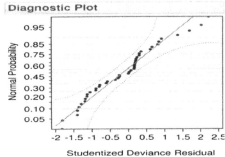

13.11   Normality seems to be satisfied and the residual plot show that the model is satisfactory.

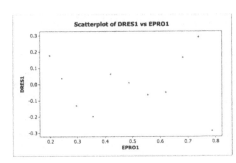
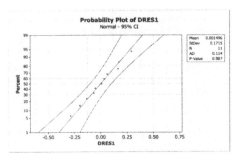

13.13 $f(y, r, \lambda) = a(\theta_1, \theta_2)b(y) \exp\{\sum c_j(\theta_1, \theta_2)d_j(y)$ gives $a(\theta_1, \theta_2) = \dfrac{\lambda^r}{\Gamma(r)}$, $b(y) = y^{-1}$, and $\sum c_j(\theta_1, \theta_2) = -\lambda y + r \ln(y)$.

13.15   Another way to write the exponential family is

$$f(y; \theta) = B(\theta)e^{Q(\theta)\,R(y)}h(y)$$

For the negative binomial, if replace $(1 - \pi)^y$ by $e^{\log(1-\pi)y}$ we get

$B(\theta)$	$Q(\theta)$	$R(y)$	$h(y)$
π^α	$\log(1 - \pi)$	y	$\dbinom{y + \alpha - 1}{\alpha - 1}$

13.17 There is no need to rework the problem since all of the regressor were important.

13.19 Both plots look good and indicate the model is adequate.

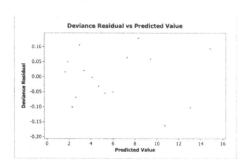
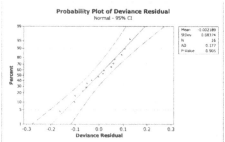

13.21 Look at

$$\widehat{\eta}(x_1 + 1) - \widehat{\eta}(x_1) = \widehat{\beta}_0 + \widehat{\beta}_1(x_1 + 1) + \widehat{\beta}_2 x_2 + \widehat{\beta}_{12}(x_1 + 1)(x_2)$$
$$- (\widehat{\beta}_0 + \widehat{\beta}_1(x_1) + \widehat{\beta}_2 x_2 + \widehat{\beta}_{12} x_1 x_2)$$
$$= \widehat{\beta}_1 + \widehat{\beta}_{12} x_2$$

Therefore, $\widehat{O}_R = e^{\widehat{\beta}_1 + \widehat{\beta}_{12} x_2}$ which includes the estimated interaction coefficient and x_2 has to be fixed.

13.23 The logit model from Problem 14.5 is $\widehat{\pi} = \dfrac{1}{1 + e^{(7.047 - 0.00007x_1 - 0.9879x_2)}}$.

G = 6.644 with p-value = 0.036, D = 18.3089 with p-value = 0.306.
The probit function is $\widehat{\pi} = \dfrac{1}{1 + e^{(4.350 - 0.000046x_1 - 0.6099x_2)}}$.
G = 6.771 with p-value = 0.034, D = 18.1819 with p-value = 0.313.
The complimentary log-log model is $\widehat{\pi} = \dfrac{1}{1 + e^{(5.737 - 0.000057x_1 - 0.7219x_2)}}$.
G = 6.871 with p-value = 0.032, D = 18.0827 with p-value = 0.319.

The likelihood ratio tests all show model significance for the three links. Also, the goodness of fit tests using the deviance show the models are very similar. This is to be expected since for small sample sizes, the three models do not show meaningful differences.

13.25 a. Using the logit function, $\widehat{\pi} = \dfrac{1}{1 + e^{(-10.875 + 0.1713x_1)}}$.

The model fits the data well.

G=5.944 with a p-value of 0.015 and D=15.7592 with p-value=0.398.

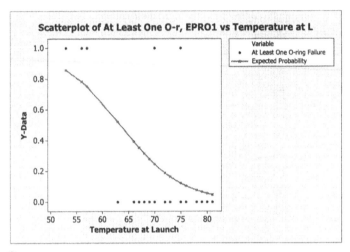

b. $\widehat{O_R} = 0.84$ This implies that an additional degree (Fahrenheit) of temperature decreases the odds of o-ring failure by 16%.

c. $\widehat{\pi} = 0.9097$ at 50 °F.

d. $\widehat{\pi} = 0.1221$ at 75 °F.

e. $\widehat{\pi} = 0.9962$ at 31 °F. There is danger in extrapolating beyond the range of temperatures used in the model, but we can see from the graph of estimated probabilities and from the calculated values in parts c and d that the probability of failure at this low temperature is very high.

f. The deviance residuals are shown below.

Temperature (F)	Deviance Residual
53	0.35569
56	0.56743
57	0.65666
63	1.60629
66	1.05038
67	3.00090
68	0.78648

Temperature (F)	Deviance Residual
69	0.67896
70	5.99277
72	0.43192
73	0.36997
75	2.09057
76	0.47858
79	0.14041
80	0.11883
81	0.10046

There may be some problems with the model.

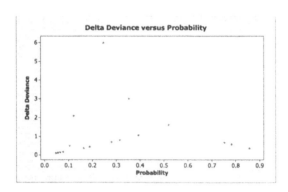

g. Using the logit function,
$$\widehat{\pi} = \frac{1}{1 + e^{(-39.1593 + 1.01923x_1 - 0.00630x_1^2)}}.$$ $G = 6.386$ with $p = 0.041$.
D $= 15.3177$ with $p = 0.357$. The plot of deviance residuals for this model looks better than that for the model in part a., suggesting this model may be an improvement to the original.

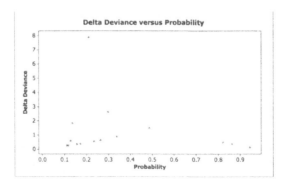

13.27 Four indicator variables were used to incorporate the five levels of dose into the analysis. A Poisson regression model with a log link was used to determine the effect of dose on the number of offspring. The model adequacy checks based on deviance ($\chi^2 = 47.44$) and the Pearson chi-square ($\chi^2 = 50.7188$) statistics are satisfactory. From the analysis, we notice when comparing to the control, dosages 235 and 310 have a significant effect on number of offspring.

Source	Test Statistic	p-value
80	0.0016	0.9682
160	1.61	0.2044
235	42.10	< 0.0001
310	189.07	< 0.0001

The residual plots show some problems with normality and model fit.

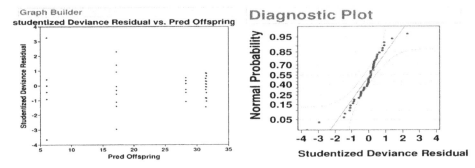

13.29 A regression with a gamma response distribution and a log link function was performed on the resistivity of a urea formaldehyde resin data. For the full model, the scaled deviance is 32.97 indicating that the model is adequate. The LR statistics for the Type III analysis indicate that some of the regressors should be removed from the model because they are not significant. Insignificant regressors were removed from the model and the resulting model only has E, the water collection time as the single predictor. The same analysis was completed using the canonical link but this had no effect on the conclusions for the analysis.

Source	Test Statistic	p-value
E	3.83	0.0503

Normality seems to be satisfied and the residual plot shows the model is satisfactory.

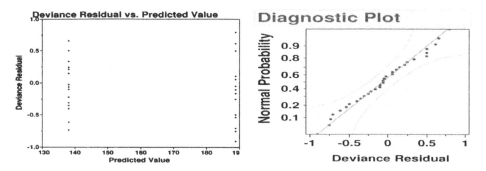

13.31 a) The logistic regression with the logit link function results in estimates $\widehat{\beta}_0 = -5.34$ and $\widehat{\beta}_1 = 0.001548$. The regressor pressure is significant with a test statistics of $\chi^2 = 112.46$ and a p-value=0.000. The Deviance=0.37 with a p-value=1.000 indicating that the model is adequate. The residual analysis does not indicate any problems with model adequacy.

b) The logistic regression with the probit link function results in estimates $\widehat{\beta}_0 = -3.271$ and $\widehat{\beta}_1 = 0.000949$. The regressor pressure is significant with a test statistics of $\chi^2 = 112.39$ and a p-value=0.000. The Deviance=0.44 with a p-value=1.000 indicating that the model is adequate. The residual analysis does not indicate any problems with model adequacy.

c) The logistic regressions with both the logit link and probit link result in very similar results with adequate fits to the data. The decision on the model to use would be left to the subject matter expert on which model has the better interpretation.

d) The Poisson regression of Failures versus Pressure with the natural log link function results in estimates $\widehat{\beta}_0 = 0.858$ and $\widehat{\beta}_1 = 0.000755$. The regressor Pressure is significant with a test statistics of $\chi^2 = 59.94$ and a p-value=0.000. The Deviance=25.05 with a p-value=0.002 indicating that the model is not adequate. The residual analysis also shows some problems with the normality assumption. The fit for using the logistic regression is preferred over the Poisson regression fit.

13.33 This is a very rich dataset that could be used to answer a variety of research questions. I chose to use binary logistic regression to answer one research question on whether the prevalence of patients with antigen levels over 35 Units/mL differed among benign diseases versus the other patients in the dataset (pancreatic cancer, lung cancer, breast cancer, colorectal cancer, gastrointestinal, nongastrointestinal, ovarian cancer). My data table is given below:

Group	Frequency	Response
benign diseases	9	yes
benign diseases	134	no
Disease	140	yes
Disease	161	no

The logistic regression with the logit link function results in estimates $\widehat{\beta}_0 = -2.701$ and $\widehat{\beta}_1 = 2.561$. The regressor Group is significant with a test statistics of $\chi^2 = 83.59$ and a p-value=0.000 revealing that prevalence of patients with antigen levels > 35 units/mL differs for patients with benign diseases versus other patients with other diseases. The Deviance=483.01 with a p-value=0.087 indicating that the model is adequate.

Chapter 14: Regression Analysis of Time Series Data

14.1   The simple linear regression model is $\widehat{y} = -3.03 + 0.0095x$. The residual plot versus time indicates there is autocorrelation. The Durbin-Watson statistic is $d = 1.2257$. There is evidence of positive autocorrelation.

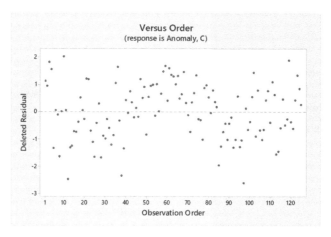

Using the Cochrane-Orcutt method we obtain $\widehat{\varphi} = 0.3819$. The new regression equation is $\widehat{y}' = -1.901 + 0.0097x$. The Durbin-Watson statistic is $d = 1.94736$. We conclude there is no problem with autocorrelated errors in the transformed model.

14.3   a. $\widehat{y} = 24.6 - 0.0892x$. The residual plot versus time indicates there is autocorrelation.

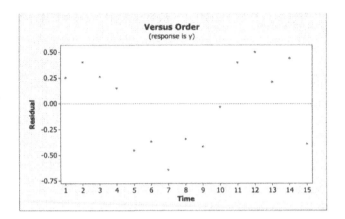

b. $d = .81$ which rejects the null hypothesis and indicates that there is evidence of positive autocorrelation.

c. We get

$$\widehat{\rho} = \frac{\sum e_t e_{t-1}}{\sum e_t^2}$$

$$= \frac{1.1693}{2.1610}$$

$$= .5411.$$

The new regression equation is $\widehat{y'} = 12.0854 - 0.1105 x'$
The standard errors of the regression coefficients are $se(\widehat{\beta_0'}) = 0.5542$ and $se(\widehat{\beta_1'}) = 0.0.01403$.

d. $d = .90$ which indicates there is still evidence of positive autocorrelation.

14.5 The regression through the origin for the first difference approach yields an estimated slope of 0.28943 with a standard error of 0.02508. The previous estimate for β_1 was 0.29799 with a standard error of 0.0123. As a result, the estimates are very similar, but the standard error is smaller for the Cochrane-Orcutt approach in exercise 14.4

14.7 The objective function is

$$\sum_{t=1}^{T} \frac{1}{t} [y_t - \widehat{\beta_0} - \widehat{\beta_1} t]^2.$$

Taking the derivatives with respect to $\widehat{\beta_0}$ and $\widehat{\beta_1}$ and setting equal to 0, we obtain

$$2 \sum_{t=1}^{T} \frac{1}{t} [y_t - \widehat{\beta_0} - \widehat{\beta_1}](-1) = 0$$

and

$$2 \sum_{t=1}^{T} \frac{1}{t} [y_t - \widehat{\beta}_0 - \widehat{\beta}_1](-t) = 0$$

$$2 \sum_{t=1}^{T} [y_t - \widehat{\beta}_0 - \widehat{\beta}_1](-1) = 0$$

The resulting normal equations are

$$\widehat{\beta}_0 \sum_{t=1}^{T} \frac{1}{t} + \widehat{\beta}_1 = \sum_{t=1}^{T} \frac{y_t}{t}$$

and

$$T(\widehat{\beta}_0 + \widehat{\beta}_1) = \sum_{t=1}^{T} y_t.$$

noindent Let

$$H_t = \sum_{t=1}^{T} \frac{1}{t}.$$

We note that H_T is the *harmonic number* represented by the partial sum through T terms. The resulting solutions to the normal equations are

$$\widehat{\beta}_0 = \frac{\sum_{t=1}^{T} \frac{y_t}{t} - \overline{y}}{H_T - 1}$$

and

$$\widehat{\beta}_1 = \overline{y} - \widehat{\beta}_0.$$

14.9   The regression equation using the Cochran-Orcutt procedure in exercise 14.3 is $\widehat{y}' = 12.0854 - 0.1105x'$ and the standard errors of the regression coefficients are $se(\widehat{\beta}_0') = 0.5542$ and $se(\widehat{\beta}_1') = 0.01403$.

The regression equation using the Cochran-Orcutt procedure in exercise 14.3 is $\widehat{y}' = 12.0854 - 0.1105x'$ and the standard errors of the regression coefficients are $se(\widehat{\beta}_0') = 0.5542$ and $se(\widehat{\beta}_1') = 0.0.01403$.

The time series regression model with autocorrelated errors produces the regression equation $\widehat{y} = 26.1875 - 0.1075x$ and the standard errors of the regression coefficients are $se(\widehat{\beta}_0) = 1.1827$ and $se(\widehat{\beta}_1) = 0.0131$.

The estimates are very similar but the standard error is smaller for the time series regression.

14.11 The time series regression has $\widehat{\phi} = 0.600189$, which is the same as the Cochrane-Orcutt procedure from before. The previous estimate for $\widehat{\beta}_1$ was 0.29799 with a standard error of 0.01230. The time series regression estimate is 0.2910 with a standard error of .009776. As a result, the estimates are very similar, but the standard error is smaller for the time series regression.

14.15 The model in Eq. (14.26) was fit to the chemical process data. The Durbin-Watson Statistic is $d = 0.95$, which still indicates a problem with autocorrelations in the residuals. The t-statistic for the lagged predictor temperature has a p-value of 0.06. The model in Eq. (14.27) was fit to the data. There is still a problem with autocorrelation with a Durbin-Watson Statistic of $d = 1.226$. The lagged predictors have not corrected the autocorrelation in the model. It is important to note, that these lagged models lead to at least one observation with an unusually high deleted residual. This residual should be investigated further.

14.17 When you have error associated with the predictor variable, this error will also translate to the prediction error of the variable of interest. In this case, a 10% lower prediction would not be expected to have too much of an impact because this error is not too large and the temperature predictor is already a rather noisy variable to begin with.

Chapter 15: Other Topics in the Use of Regression Analysis

15.1 It is possible, especially in small data sets, that a few outliers that follow the pattern of the "good" points can throw the fit off.

15.3 They are both oscillating functions that have similar shapes with Tukey's bi-weight being a faster wave. However, Tukey's bi-weight can exceed 1 while Andrew's wave function cannot.

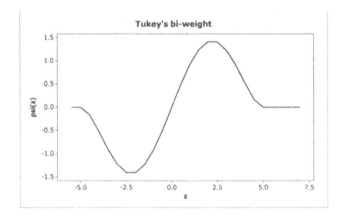

15.5 The fitted model is $\widehat{y} = 2.34 - .288x_1 + .248x_2 + .45x_3 - .543x_4 + .005x_5$ with a couple of outliers.

15.7 a. The estimate is

$$\widehat{x}_0 = \frac{y_0 - \widehat{\beta}_0}{\widehat{\beta}_1}$$

$$= \frac{17 - 33.7}{-.0474}$$

$$= 352.32$$

b. First we solve the following

$$d^2 \left[(-.0474)^2 - \frac{(2.042)^2(9.39)}{426101.75} \right] - 2d(-.0474)(17 - 20.223)$$

$$+ \left[(17 - 20.223)^2 - (2.042)^2(9.39)\left(1 + \frac{1}{32}\right) \right] = 0$$

$$.022d^2 - 3.055d - 29.99 = 0$$

which gives $d_1 = -66.41$ and $d_2 = 205.27$. Then the confidence interval is $285.04 - 66.41 < x_0 < 285.04 + 205.27$

$$218.63 < x_0 < 490.31$$

15.9   The normal-theory confidence interval for β_2 is $.014385 \pm 1.717(.003613) = (.0082, .0206)$. The bootstrap confidence interval is $(.0073, .0240)$ which is similar to the normal-theory interval.

15.11 First, fit the model. Then, estimate the mean response at x_0. Bootstrap this m times and store all of these mean responses. Finally, find the standard deviation of these responses.

15.15 Regression tree for NFL data:

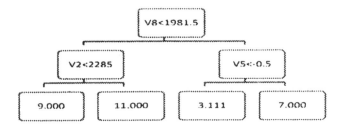

15.17 $Var(\widehat{\beta}_0) = \sigma^2 \left(\dfrac{1}{n} + \dfrac{\overline{x}}{S_{XX}} \right)$ which for fixed n is minimized when $\overline{x} = 0$. If this is not possible, then the experimenter should maximize S_{XX}.

15.19 a. Let $\mathbf{D}$ be the $\mathbf{X}$-matrix without the intercept column. Then $\mathbf{D} = (\mathbf{d}_1 \ \mathbf{d}_2 \ \cdots \ \mathbf{d}_k)$. Suppose the spread of the design is bounded (it has to be) then, $\mathbf{d}'_i \mathbf{d}_i \le c_i^2$ for $i = 1, 2, \ldots k$ and some constant c_i. This is equivalent to

$$d_{ii} \le c_i^2 \quad i = 1, 2, \ldots, k$$

where d_{ii} is the ith diagonal element of $\mathbf{D}'\mathbf{D}$. It can be shown that

$$d^{ii} \ge \frac{1}{d_{ii}} \quad i = 1, 2, \ldots, k$$

where d^{ii} is the ith diagonal element of $(\mathbf{D'D})^{-1}$. There is equality in the above expression only when all the d_{ij}'s $= 0$. Therefore, if the design is orthogonal

$$Var(\widehat{\beta}_i) = \sigma^2 d^{ii}$$

$$\geq \frac{\sigma^2}{c_i^2}$$

$$= \frac{\sigma^2}{c_i^2}$$

since $\mathbf{d}_i'\mathbf{d}_j = 0$ for $i \neq j$ and $\mathbf{d}_i'\mathbf{d}_i = c_i^2$ when the design is orthogonal.

b. $Var(\widehat{y}) = (\sigma^2)\mathbf{x}_0'(\mathbf{X'X})^{-1}\mathbf{x}_0$. Since the design is orthogonal, we have

$$(\mathbf{X'X})^{-1} = \begin{pmatrix} \frac{1}{n} & 0 & 0 \\ 0 & \frac{1}{n} & 0 \\ 0 & 0 & \frac{1}{n} \end{pmatrix}$$

Consider the center of the design as $\mathbf{0}$, then for any $\mathbf{x}_0 = \begin{pmatrix} 1 & x_i & x_j \end{pmatrix}$ it has distance from the center of $d = \sqrt{1 + x_i^2 + x_j^2}$ and

$$Var(\widehat{y}) = \left(\frac{1}{n}\right)^2 + \frac{x_i^2}{n^2} + \frac{x_j^2}{n^2}$$

$$= \frac{1}{n^2}(1 + x_i^2 + x_j^2)$$

$$= \frac{d^2}{n^2}$$

Thus, for any point with distance d the variance will be the same which means the design is rotatable.

15.21 Runs at the center of the design region allow for an estimate of pure error. However, you can also get an estimate of pure error by replicating factorial runs. What you gain is a test of lack of fit. By putting all four runs at the center, you maximize the power of this lack of fit test.

15.23 a)

Run	X1	X2
1	0.75	1
2	1	1
3	0.5	1
4	1	1
5	1	0.5
6	0.5	1
7	1	0.75
8	1	0.5
9	0.75	1
10	1	0.75
11	0.75	0.75
12	0.75	0.75

b)

Run	X1	X2
1	0.5	1
2	1	0.7376694805
3	1	1
4	0.75	1
5	0.9375057418	0.780871143
6	1	0.7385276615
7	0.75	0.75
8	0.75	0.75
9	1	0.5
10	0.7423928648	1
11	0.7863954606	0.9315035011
12	1	1

c) In terms of estimation efficiency, the relative standard error of the regression coefficient estimates can be compared for the two designs. The relative standard errors of the estimates are slightly lower for the I-optimal design compared to the D-optimal design, which is expected. The D-optimal design has an average predication variance of 0.3182 σ^2 and 0.2833 σ^2 for the I-optimal design. This is not surprising because the I-optimal design is constructed to minimize this quantity.

Printed and bound by CPI Group (UK) Ltd, Croydon, CR0 4YY

27/10/2024

14580280-0005